SpringerBriefs in Stem Cells

For further volumes:
http://www.springer.com/series/10206

Tiziana A. L. Brevini · Georgia Pennarossa

Gametogenesis, Early Embryo Development and Stem Cell Derivation

 Springer

Tiziana A. L. Brevini
Laboratory of Biomedical Embryology
Università degli Studi di Milano—Unistem
Milan
Italy

Georgia Pennarossa
Laboratory of Biomedical Embryology
Università degli Studi di Milano—Unistem
Milan
Italy

ISSN 2192-8118
ISBN 978-1-4614-5531-8
DOI 10.1007/978-1-4614-5532-5
Springer New York Heidelberg Dordrecht London

ISSN 2192-8126   (electronic)
ISBN 978-1-4614-5532-5   (eBook)

Library of Congress Control Number: 2012948637

Printed on acid-free paper

Springer is part of Springer Science+Business Media (www.springer.com)

Foreword

Science is evolving at an immense speed. Breakthroughs, which most of us had never even thought of few years ago because they clashed against cemented dogmas, are now realities. They now pave the way for further scientific endeavors and offer new solutions to important societal problems. In the area of biology, some of the greatest breakthroughs over the past years have been the birth of the cloned sheep, Dolly, in 1996, and the finding just 10 years later in 2006 that differentiated somatic cells can be reprogrammed and revert to a pluripotent stem cell state. The two achievements have commonalities: Both have demolished the biological dogma that terminally differentiated cells cannot de-differentiate, both are based upon a total reprogramming of the epigenetic control of cellular gene expression, and both have their roots in embryology. It is embryological thinking that is the background for understanding how an oocyte can be used for reprogramming of a somatic cell for the creation of a cloned animal, and it is embryological reasoning that resulted in isolation and culture of embryonic stem cells and, later, the identification of exclusive stem cell factors of such a potency that they can reprogram somatic cells into stem cells. As visualized by just these two examples, embryology stands out as a modern contemporary scientific discipline, in spite of its classical nature with a history dating all the way back to Aristotele (384–322 BC).

The immense speed by which science evolves not only reflects in spectacular breakthroughs, as alluded to above, it is also reflected in the level and depth of molecular understanding of processes. A wealth of data are generated that allows us to penetrate deeper and deeper into the molecular understanding of how life is organized, controlled, and passed from generation to generation. The mapping of the genomes of different organisms has contributed significantly to this process, but over the past years an enormously focussed penetration into the understanding of the epigenetic landscape, which controls gene expression and silencing, has been instrumental. And again, embryology is central in this aspect: The full understanding of epigenetic reprogramming can exclusively be obtained if this phenomenon is seen in an embryological perspective. Hence, the understanding of the epigenetic mechanisms, which operate when the genome is passed from one generation to the next, is crucial for normal reproduction as well as for contemporary phenomena like fetal programming, where there is a growing body of evidence that the mother affects the epigenetic patterns of the embryo and fetus to such a degree that it is decisive for the rest of the life. It's all embryology …

The present book stands out as a contemporary appreciation of the most important aspects of embryology that are important to grasp not only in relation to stem cell biology, but also as a background for assisted reproductive technologies and several other scientific methodologies. Tiziana Brevini is an internationally highly recognized and respected embryologist who has contributed to numerous aspects of the area, from oocyte biology to stem cell culture and differentiation. It is a gift to the scientific area of embryology that she has devoted time for the book in your hands.

Prof. Poul Hyttel
Department of Veterinary Clinical and Animal Sciences
University of Copenhagen
Copenhagen
Denmark

Preface

In a somehow limited view of this discipline, Embryology has been considered in the past years a prerequisite and a fundamental acquisition for a better and more dynamic understanding of gross anatomy. We are certainly not denying this idea that has a solid ground and highlights the impact of the complex differentiation processes in the definition of the final architectural morphology of a tissue/organ and the related function.

At present, however, we are convinced that this view needs to be expanded considering the central role played by Embryology in a series of new scientific fields. Innovative and quickly developing research in biomedical science and modeling find solid bases in recently acquired information related to embryo induction and differentiation. Similarly, the latest exciting scientific acquisitions in stem cell research and regenerative medicine have been supported by the elucidation of the mechanisms and molecules controlling pluripotency and driving commitment and differentiation in the early embryo.

This Brief is intended as a concise, handy overview of the main concepts related to Embryology, re-visited through the novel concepts that are applied daily in stem cell research and cell therapy oriented investigations.

Milan, Italy

Tiziana A. L. Brevini
Georgia Pennarossa

Acknowledgments

Several qualified and enthusiastic scientists have dedicated time and made precious comments over the text and concepts presented in this Brief. Fulvio Gandolfi, and Sara Maffei, University of Milan and Josè Roberto Silva, Federal University of Ceara, were engaged in daily base discussions and were an encouragement with their criticism and appreciations. Luca Passello was a great help in the preparation of images and tables.

A special thanks to Cecilia Gandolfi for critical discussion and editing of the text, and to the students of our Anatomy and Embryology classes. Their direct questions, never-ending curiosity and enthusiasm, warmed the long hours spent around the steel tables of the dissection room and provided a refreshing, challenging cue for our teaching tasks.

Contents

Contents

Chapter 1
Gametogenesis

1.1 Primordial Germ Cells

Early embryo cells have the capability to give rise to all cell types of the body. During the process of gastrulation, however, most of them lose this ability and acquire a tissue specific fate. This loss of pluripotency is a key event in development since it results in lineage commitment and allows the definition of the three somatic germ layers from which all different tissues of the body will originate. Through a series of complex and finely orchestrated morphogenetic movements, that involve cell migration, clustering and delamination, the process leads to the formation of the ectoderm, the mesoderm, and the endoderm (Fig. 1.1).

During the process of gastrulation, however, a subpopulation of cells remains pluripotent and does not undergo any lineage commitment. These cells are the common progenitors of the male and female gametes of the developing embryo and are known as the primordial germ cells. Like all other somatic cells they are diploid and, in human embryos, they can be identified in the epiblast or primary ectoderm as early as the second week of gestation. Shortly after that, the primordial germ cells migrate from the primary ectoderm into the yolk sac wall and localize outside the embryo proper. It is believed that the primordial germ cells are moved to this location outside the developing embryo in order to release them from the differentiation cues driving gastrulation and patterning (Fig. 1.2). This seclusion would ensure them a neutral environment to preserve their pluripotency.

During these stages, primordial germ cells can be identified by their relatively big dimensions and through immunostaining carried out using antibodies specific for transcription factors such as Oct 4, Stella, VASA, Fragilis, BLIMP-1, and Alkaline phosphatase (see Table 1.1).

Recent studies have demonstrated that some of these factors are directly related to the pluripotent state and are actively repressing the expression of key genes driving differentiation of epiblast cells to the somatic lineages, thus concurring in maintaining a "non-committed" environment for primordial germ cells.

T. A. L. Brevini and G. Pennarossa, *Gametogenesis, Early Embryo Development, and Stem Cell Derivation*, SpringerBriefs in Stem Cells, DOI: 10.1007/978-1-4614-5532-5_1, © The Author(s) 2013

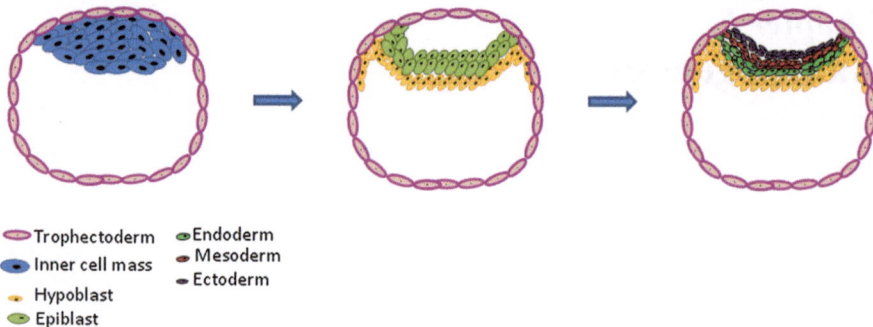

- Trophectoderm   - Endoderm
- Inner cell mass - Mesoderm
- Hypoblast   - Ectoderm
- Epiblast

Fig. 1.1 Formation of the three germ layers. The blastocyst consists of two cell types: trophecto-derm (placenta formation) and inner cell mass (embryo proper) cells. At the end of blastulation, the inner cell mass cells give rise to hypoblast (internal) and epiblast (external) cells, establishing the bilaminar embryonic disk. During the process of gastrulation, epiblast cells differentiate into ecto-derm, mesoderm, and endoderm, converting the bilaminar disk into a trilaminar embryonic disk

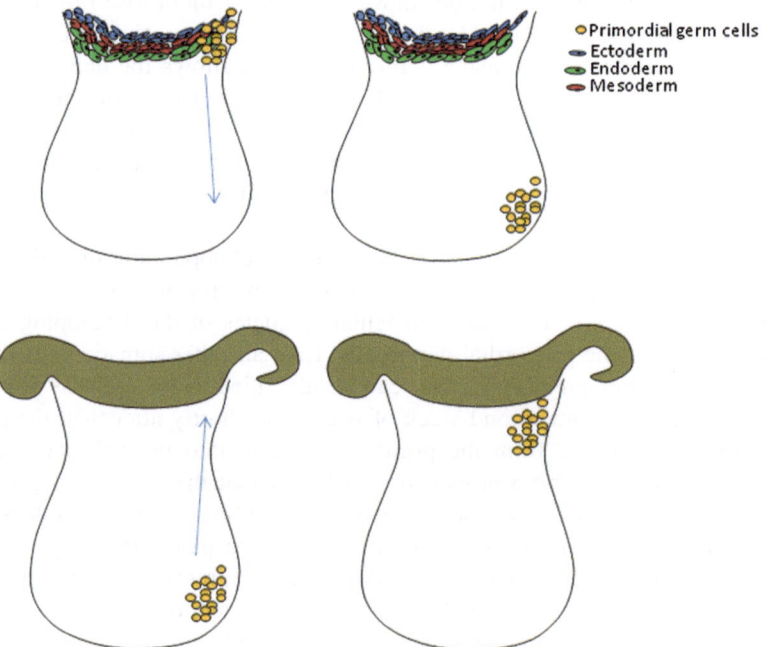

- Primordial germ cells
- Ectoderm
- Endoderm
- Mesoderm

Fig. 1.2 Primordial germ cell migration. During the process of mesoderm and endoderm forma-tion, the primordial germ cells move from the primary ectoderm into the yolk sac wall and local-ize outside the forming embryo. Here these cells proliferate actively, without differentiating, and then migrate again to colonize the genital ridge of the embryo

Soon, primordial germ cells start migrating along the caudal region of the yolk sac and wander back into the embryo where they colonize the gonadal ridge. It is yet unclear whether this relocation is actively or passively controlled but possibly

Table 1.1 Pluripotency-related transcription factors identified in primordial germ cells

Gene symbol	Gene name
Akp2	Alkaline phosphatase, liver/bone/kidney
Blimp1	PR domain-containing 1, with ZNF domain
BMP4	Bone morphogenetic protein 4
c-kit	Kit oncogene
c-myc	Myelocytomatosis oncogene
Dax1	Dosage-sensitive sex reversal, adrenal hypo-plasia critical region, on chromosome X, gene 1
Dmc1	DMC1 dosage suppressor of mck1 homolog, meiosis-specific homologous recombination (yeast)
Figα	Factor In the Germline alpha
Fragilis	Interferon-induced transmembrane protein 3
Fgf9	Fibroblast growth factor 9
miRNAs	Different microRNA cluster
Nanog	Nanog homeobox
Nanos3	Nanos homolog 3, Drosophila
Oct4	POU domain, class 5, transcription factor 1
Scp3	Synaptonemal complex protein 3
Sox2	Sry-box containing gene 2
Sox9	Sry-box containing gene 9
Sry	Sex determining region of chromosome Y
Ssea-1	Fucosyltransferase 4
Stella	Developmental pluripotency-associated 3
Stra8	Stimulated by retinoic acid gene 8

it may be facilitated by the rostral-caudal curvature and the folding of the embryo. At present, it is widely accepted that local signals, generating outside the embryo proper, act through the Smad pathway, and involve the Bone Morphogenetic Protein 4, and 8b. Together with the many factors described above and summarized in Table 1.1, these molecules drive the germ line specification process and growth. Indeed during their journey, but also once in the gonadal ridge, primordial germ cells increase in number and multiply by mitotic divisions. These proliferation events are finely tuned and, in the mouse, eight proliferation cycles of 16 h each, with an increase in cells from about 100 to 20,000, have been reported. Furthermore, the involvement of growth factors that are directly stimulating cell proliferation and molecules like kit, KL, and LIF that down-regulate and prevent apoptosis has also been described. A few days after colonization of the genital ridge, primordial germ cells undergo mitotic arrest, associate with the surrounding somatic cells, and engage in sex-driven differentiation (Fig. 1.3). Although the sex of a mammalian embryo is genetically determined at fertilization, the genital ridges are kept in an undifferentiated state during the early phase of gestation (Figs. 1.4 and 1.5).

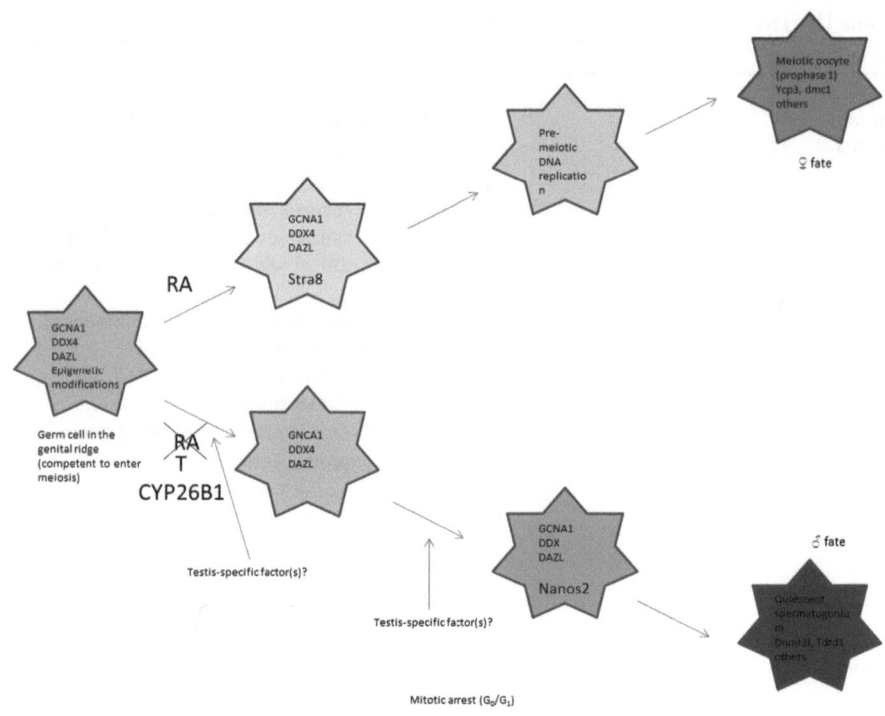

Fig. 1.3 Molecular regulation of germ line development. A complex network involving several factors regulates the acquisition of the gender-specific fate

Fig. 1.4 Scheme of
indifferent female gonad

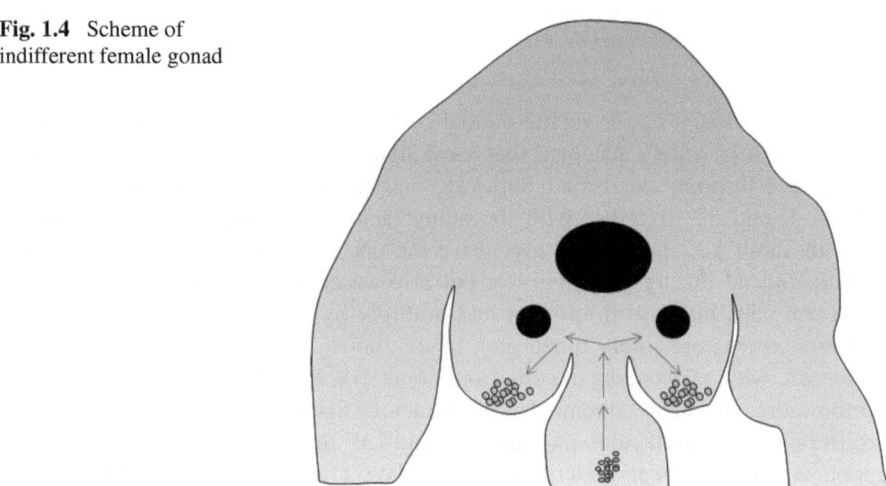

During this period, they are referred to as indifferent gonads or primitive gonadal primordium and, only following the migration of the primordial germ cells do these structures develop into the definitive and gender-specific gonads.

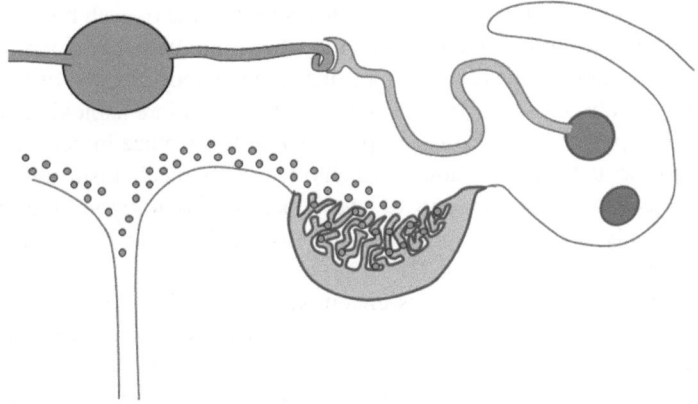

Fig. 1.5 Scheme of indifferent male gonad

Further Reading

Alberts B, Johnson A, Lewis J, Raff M, Roberts K, Walter P (2002) Primordial germ cells and sex determination in mammals. Molecular biology of the cell. 4th edn. Garland Science, New York

Bao S, Leitch HG, Gillich A, Nichols J, Tang F, Kim S, Lee C, Zwaka T, Li X, Surani MA (2012 Jul 6) The germ cell determinant blimp1 is not required for derivation of pluripotent stem cells. Cell Stem Cell 11(1):110–117

Bowles J, Koopman P (2010) Sex determination in mammalian germ cells: extrinsic versus intrinsic factors. Reproduction 139:943–958

Herpin A, Cunningham C (2007) Cross-talk between the bone morphogenetic protein pathway and other major signaling pathways results in tightly regulated cell-specific outcomes. FEBS J 274(12):2977–2985

Soto-Suazo M, Zorn TM (2005) Primordial germ cells migration: morphological and molecular aspects. Anim Reprod 2(3):147–160

Sabour D, Araúzo-Bravo MJ, Hübner K, Ko K, Greber B, Gentile L, Stehling M, Schöler HR (2011 Jan 1) Identification of genes specific to mouse primordial germ cells through dynamic global gene expression. Hum Mol Genet 20(1):115–25

1.2 Mitosis and Meiosis

Hereditary traits are determined by specific DNA segments, identified as "genes". Together with several proteins, DNA is organized in chromosomes that are inherited from the mother and the father. Somatic cells present the full number of chromosomes. Every cell contains two of each type of chromosome, forming the diploid chromosome complement designed as 2n. One chromosome of each homologous pair is inherited from the mother (oocytes) and the other from the father (spermatozoa), at the time of fertilization. This is possible thanks to the meiotic division of gametes that contain only one chromosome from each pair, and are therefore haploid (1n).

Meiosis is a particular cell division process that takes place in germ cells and is necessary for sexual reproduction in eukaryotes. This mechanism allows obtaining haploid cells, containing one of every pair of homologous chromosomes, and involves only one DNA replication and two distinct and consecutive nuclear reductions (meiosis I and meiosis II).

In particular, the meiotic process encompasses interphase, meiosis I and meiosis II. The interphase consists of G1 and S phases, while meiosis I and II are divided into prophase, metaphase, anaphase, and telophase stages, similarly to the corresponding phases in the mitotic cell cycle. The meiotic process therefore includes the stages of meiosis I (prophase I, metaphase I, anaphase I, telophase I), and meiosis II (prophase II, metaphase II, anaphase II, telophase II).

During interphase gametes synthesizes a vast array of proteins, including enzymes and structural proteins required for growth (G1 phase), and in order to duplicate each of the chromosomes forming a complex of two identical sister chromatids (S phase).

This is followed by the first step of the prophase of meiosis I, the leptotene stage, wherein the chromosomes condense and align in pairs along the center of the nucleus. Each chromosome therefore consists of four chromatids and is referred to as a tetrad, in which maternal chromatids become bound to its paternal counterpart by the synaptonemal complex (zygotene stage). During the pachytene stage, this coupling allows for DNA to be exchanged between homologous chromatid segments at certain points—called chiasmata—through a crossover process, resulting in a new DNA combination. This event, together with the random segregation of maternal and paternal chromosomes, represents a significant source of genetic recombination and ensures a unique combination of parental genomes for the next generations.

During the following phases, diplotene and diakinesis, the synaptonemal complex degrades, homologous chromosomes are gradually released from each other and further condense, the nucleolus disappears together with the nuclear membrane, and the mitotic spindle begins to form. Furthermore in male gametes, with the exception of the mouse, centriole pairs, which were duplicated during S phase, migrate to opposite poles of the cell, forming microtubule organizing centers. However, in the non-rodent mammalian female, the oocyte degrades its centrioles and retains a stockpile of centrosomal proteins that are used to make up the spindle poles. The microtubules occupy the nuclear region, attaching to the chromosomes at the kinetochore. During metaphase I, homologous chromosomes align along the equatorial plane; in anaphase I, kinetochore microtubules shorten, pulling homologous chromosomes toward opposing poles, thus forming two haploid groups (telophase I). At this point, each chromosome, which still contains a pair of sister chromatids, decondenses uncoiling back into chromatin, microtubules disappear and a new nuclear membrane surrounds each haploid set.

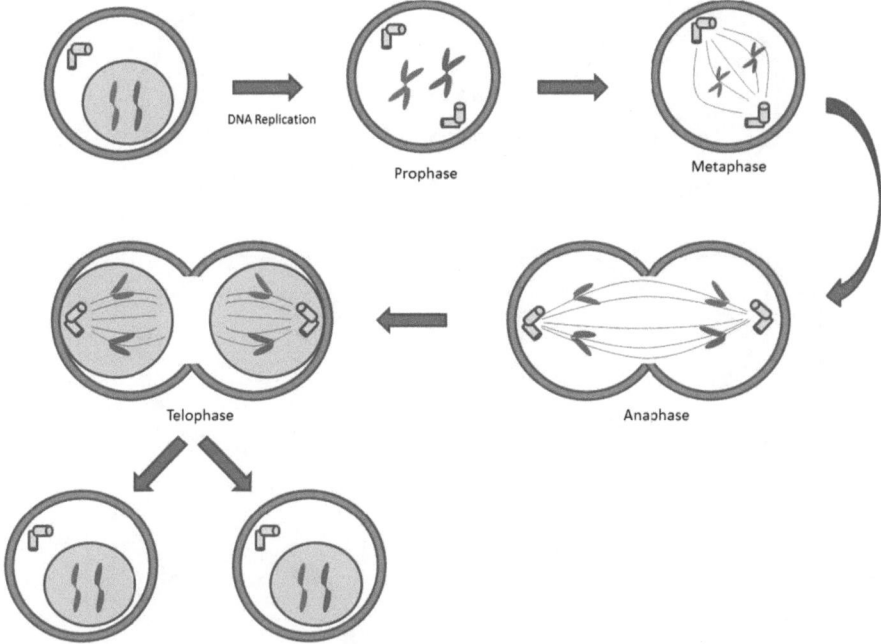

Fig. 1.6 Diagram of mitosis: the process by which one somatic cell produces two daughter cells identical to one another and to the original parent cell. Mitosis is divided into four principals stages: Prophase, Metaphase, Anaphase, and Telophase. During prophase, the chromatin condenses into chromosomes consisting of two sister chromatids. The latter then align at the equatorial plate, attaching to microtubules of the mitotic spindle (Metaphase). The sister chromatids separate and move toward opposite poles (Anaphase) and the nuclear envelope reappears (Telophase). Finally the cytoplasm divides, producing two daughter cells

At the end of meiosis I, the primary spermatocyte divides in two secondary spermatocytes, while the oocyte gives rise to one larger daughter cell (secondary oocyte) and one smaller cell without organelles (first polar body).

After completing meiosis I, the male and female gametes may begin the meiosis II process, with no DNA replication, thus going directly from telophase I to prophase II without the interphase. In prophase II, chromosomes with two chromatids again condense, nucleoli disappear, the nuclear envelope dissolves, and centrioles move to the polar regions arranging spindle fibers. The chromosomes move into the center of the cell, forming a new equatorial metaphase plate (metaphase II), next the kinetochores move toward the poles, splitting up the sister chromatids (anaphase II). The process ends with telophase II, during which the cells divide for the last time, concentrating the chromatids in the opposite poles, dissolving the spindle, reforming the nuclear envelope, and decondensing chromosomes.

At this point a total of four haploid cells, each with a half set of chromosomes, are produced. In fact, the secondary spermatocyte divides into two spermatids, while the secondary oocyte splits into one large cell (precursor of zygote) and one smaller cell (second polar body). It is important to highlight that the ovulated

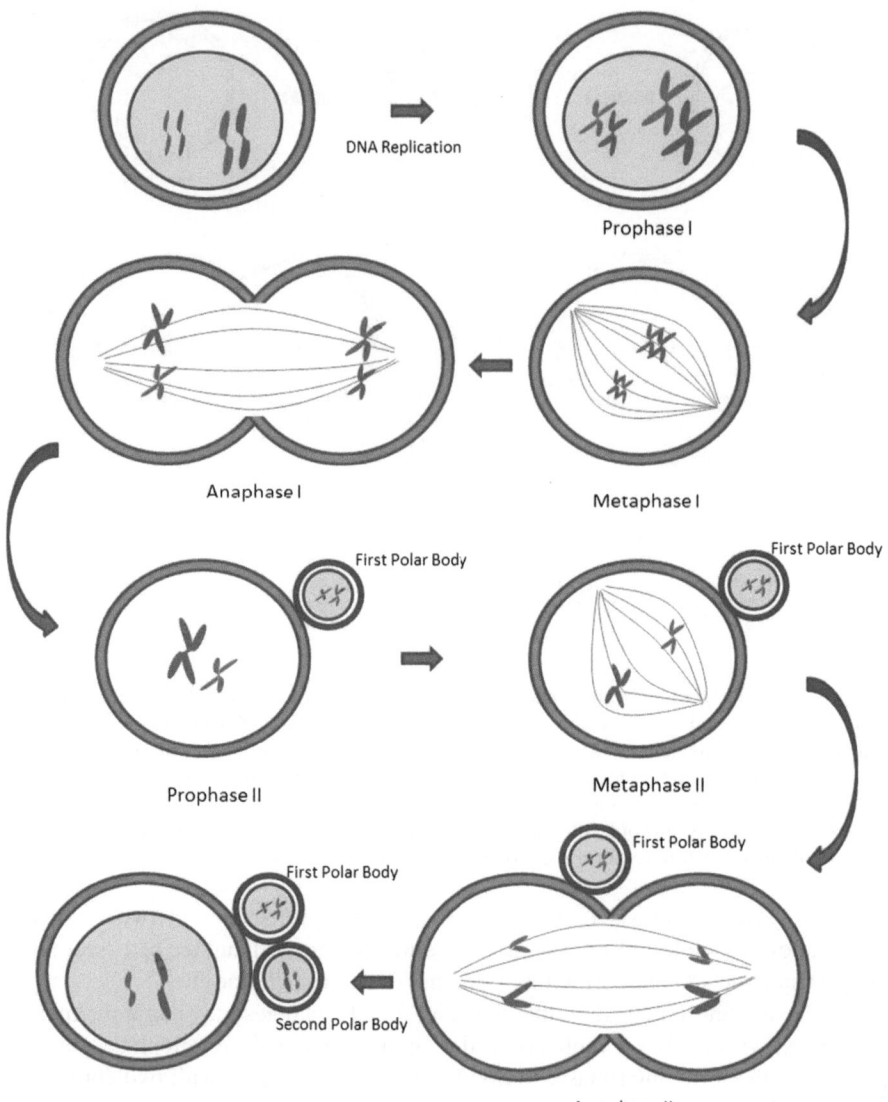

Fig. 1.7 Diagram of meiosis: the process by which a germ cell produces haploid cells, containing one of every pair of homologous chromosomes. It involves only one DNA replication and two distinct and consecutive nuclear reductions (meiosis I and meiosis II), which are divided into Prophase, Metaphase, Anaphase, and Telophase. Briefly, during prophase I the chromosomes condense. They move into the center of the cell and align in pairs along the equatorial plane, forming the tetrad, and crossover takes place (Metaphase I). During the following phases, chromosomes are pulled to opposite poles and form two haploid groups, containing a pair of sister chromatids. After completing meiosis I, gametes directly enter meiosis II, without any DNA replication. In prophase II, chromosomes with two chromatids again condense. They form a new equatorial metaphase plate (metaphase II) and then move toward the poles, splitting up the sister chromatids (anaphase II). The process ends with telophase II, during which the cells divide for the last time, producing haploid cells, thus each with a half set of chromosomes

Table 1.2 Comparison of meiosis and mitosis. Definition, steps, functions, and characteristics of the two processes

	Meiosis	Mitosis
Definition	A process in which the gamete divides into four haploid cells, reducing the number of chromosomes and separating homologous chromosomes	A process in which the cell divides into two identical diploid cells, maintaining an equal number of chromosomes
Occurs in	Human, animals, plants	All organisms
Steps	*Meiosis I*: Interphase, prophase I, metaphase I, anaphase I, and telophase I; *Meiosis II*: Prophase II, metaphase II, anaphase II, and telophase II	Interphase, prophase, metaphase, anaphase, telophase, and cytokinesis
Number of DNA replications	1	1
Number of cell divisions	2	1
Pairing of homologs	Yes	No
Crossing over	Yes	No
Number of daughter cells produced and their chromosomal assessment	4 haploid cells	2 diploid cells
Creates	Gametes only: Female egg cells or Male sperm cells	Somatic cells
Function	Sexual reproduction	Somatic cell replication

oocyte is arrested in metaphase II and only after fertilization it becomes able to conclude the meiosis II process (Figs. 1.6, 1.7 and Table 1.2).

Further Reading

Alberts B, Johnson A, Lewis J, Raff M, Roberts K, Walter P. Mitosis (2002) Molecular biology of the cell, 4th edn. Garland Science, New York

Bernstein H, Bernstein C (2010) Evolutionary origin of recombination during meiosis. Bioscience 60(7):498–505

Blow J, Tanaka T (2005) The chromosome cycle: coordinating replication and segregation: second in the cycles review series. EMBO Rep 6(11):1028–1034

Lodish H, Berk A, Zipursky L, Matsudaira P, Baltimore D, Darnell J (2000) Overview of the cell cycle and its control. Molecular cell biology. W.H Freeman, New York

Snustad DP, Simmons MJ (2008) Principles of genetics 5th edn. Wiley, New Jersey

Wilkins AS, Holliday R (2009) The evolution of meiosis from mitosis. Genetics 181(1):3–12

Fig. 1.8 Primordial follicle.
It consists of a primary
oocyte surrounded by a single
layer of squamous follicular
cells

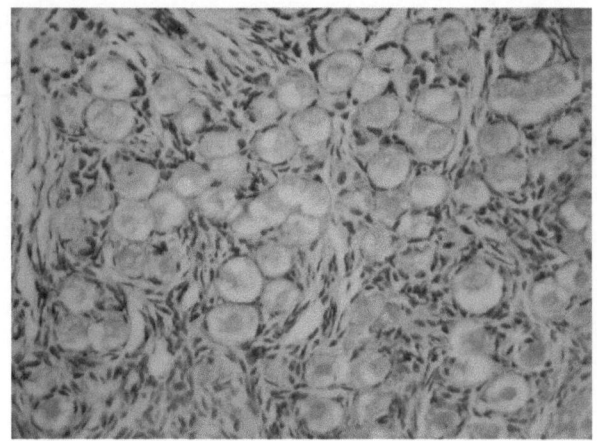

1.3 Maturation of the Female Gamete

Primordial germ cells (PGCs) move across the embryo, as previously described, to reach their specific site of function: the gonad. Here they stop migrating, associate with somatic gonadal precursor cells (SGPs) and, in the absence of the Sry gene, differentiate into eggs, thus starting to carry out their distinctive functions. In particular, in female reproductive tract, primordial germ cells give rise to oogonia which proliferate rapidly, without complete cytokinesis, or continue to differentiate into primary oocytes. Despite this, apoptotic rate is rather high during fetal development and oogonia and primary oocytes degenerate before birth. The surviving primary oocytes initiate the meiotic process, transiting through S phase and prophase I. They then stop at the diplotene stage, maintaining an intact nucleus, and remain quiescent within a protective shell of follicular cells, forming primordial follicles (Fig. 1.8).

Follicular development, also known as folliculogenesis, begins after puberty when primordial follicles are recruited to initiate growth, and subsequently develop into primary, secondary, and tertiary follicles. This process is characterized by significant morphologic and hormonal changes.

Moreover, it is important to highlight that the majority of primordial follicles that enter in a growth phase fail to complete development and degenerate thorough a process known as atresia.

Recruitment of primordial follicles seems to be mediated by different hormone stimulations and/or inhibitions as well as by a production of local growth factors.

Upon activation, follicular cells start to proliferate and change their morphology from a flat to a cuboidal structure, constituting the typical monolayer (granulosa cells) surrounding the oocyte of the primary follicle (Fig. 1.9).

At the same time, basic paracrine signaling pathways between oocyte and granulosa cells are formed and the oocyte grows dramatically, increasing its diameter from

Fig. 1.9 Primary follicle. It consists of a primary oocyte surrounded by a single layer of cuboidal follicular cells

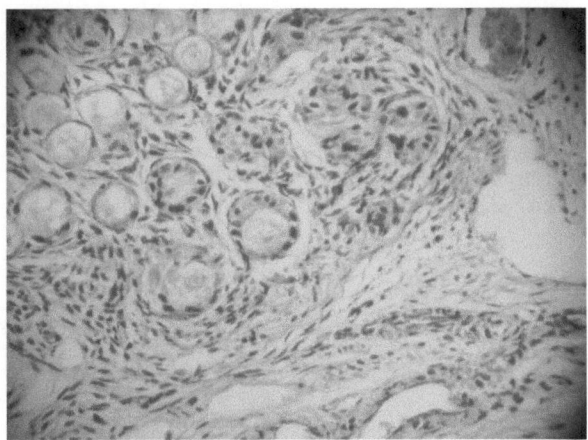

Fig. 1.10 Secondary follicle. The oocyte is surrounded by several layers of follicular cells (stratum granulosum). The zona pellucida, between the oocyte and follicular epithelium, becomes visible and the stroma organizes itself and differentiates into theca externa and interna

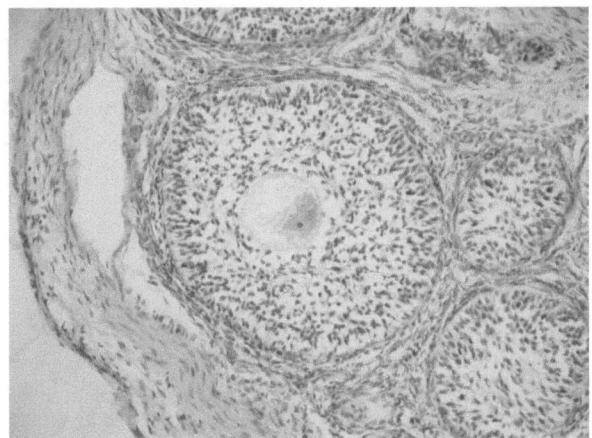

30 μm to 0.1 mm. The zona pellucida, a glycoprotein polymer capsule, becomes visible around the oocytes and granulosa cells continue to proliferate to form several layers, giving rise to the secondary follicle. Stroma cells are recruited by oocyte-secreted signals and differentiate into theca externa (supporting cells) and theca interna (steroid producing cells). Moreover, a network of capillary vessels forms between the two thecas and blood begins to circulate to and from the follicle (Fig. 1.10).

The next follicular stage, known as tertiary, antral or Graafian follicle stage, is characterized by the formation of a fluid-filled cavity adjacent to the oocyte, known as the antrum, and by movement of the oocyte into a protrusion of granulosa cells, the cumulus-oophorus cells (Fig. 1.11).

If the tertiary follicle is selected for ovulation, it enters a final step of development triggered by the increase of LH and grows quickly to become a preovulatory follicle.

Fig. 1.11 Tertiary or
Graafian follicle. It is
characterized by the presence
of a fluid-filled cavity,
the antrum. The oocyte is
surrounded by granulosa
epithelial cells, which form
the cumulus oophorus

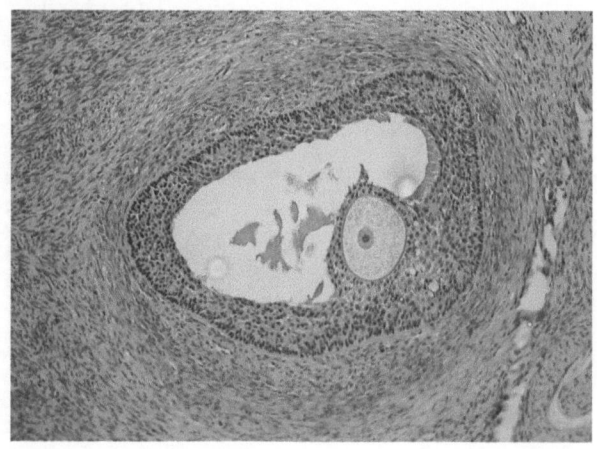

Until this last stage, the oocyte is arrested in prophase of meiosis I (germinal vesicle) and, only during the late preovulatory stage, do steroid hormones stimulate its nuclear and cytoplasmatic maturation. In particular, the oocyte resumes the meiotic process, through the breakdown of the germinal vesicle, and arrests in metaphase II, ready for ovulation (Fig. 1.12).

Further Reading

Edson MA, Nagaraja AK, Matzuk MM (2009) The mammalian ovary from genesis to revelation. Endocr Rev 30(6):624–712

Eroschenko VP (2012) DiFiore's Atlas of histology with functional correlations. Lippincott Williams & Wilkins, Philadelphia

Hyttel P, Sinowatz P, Vejlsted M, Betteridge K (2010) Essentials of domestic animal embryology. Saunders Elsevier, Philadelphia

Gilbert SF (2000) Developmental biology, 6th edn. Sinauer Associates, Sunderland

Sadler TW (2011) Langman's medical embryology, 9th edn. Lippincott Williams & Wilkins, Philadelphia

1.4 Imprinting and Epigenetic Regulation

Mammalian species inherit two copies of every gene, one comes from the mother and the other from the father. For the majority of genes both of the inherited copies are functional; a small group of genes however display a distinct regulation and one copy, either of maternal or paternal origin, is turned off. These genes are

Fig. 1.12 Representative
scheme of female gamete
maturation

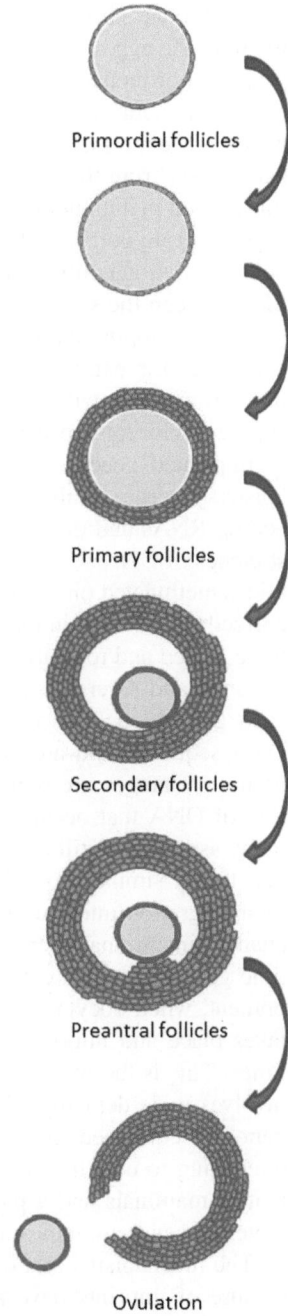

referred to as 'imprinted' because one copy of the gene was epigenetically marked in either the egg or the sperm. This indicates that mammalian development needs a proper, coordinated, and controlled expression of both the paternal and the maternal genome. In particular, it is at present accepted that, while a large number of genes are expressed from both parental alleles, about 100 genes are expressed exclusively from the maternally or paternally inherited chromosome, and therefore transcribed in a monoallelic manner (Table 1.3).

Several hypotheses have been formulated to explain imprinting and it is believed that genomic imprinting evolved in mammals because of a parental conflict between the sexes in order to regulate the maternal consumption of resources for the developing embryo. Paternally expressed imprinted genes are addressed to encourage the extraction of nutrients from the mother during gestation, therefore they promote growth. On the other hand, the maternal genome tends to limit and suppress nutrient deprivation (Fig. 1.13).

Imprinted genes are epigenetically marked and are located in well-identified clusters in order to allow them to share common regulatory elements, such as noncoding RNAs and differentially methylated regions (DMRs). The latter are usually stretches of DNA rich in cytosine and guanine nucleotides, with the cytosine nucleotides methylated on one copy but not on the other. The epigenetic marks need to be correctly positioned in the germline, they must be retained throughout life, but have to be erased and repositioned in the germline of the next generation. Imprinting is a dynamic and reversible process. Imprints must be erased and repositioned through each generation. This explains why the related modifications do not involve the DNA sequence and are mainly epigenetic. At present, we know that these reprograming changes are regulated through specific waves of methylation/demethylation of DNA that occur during gametogenesis and early embryo development but many aspects are still to be elucidated. PGCs and somatic cells initially appear epigenetically similar. Both display high levels of DNA methylation. However, during their migration into the genital ridge, the developing germ cells undergo dramatic changes in chromatin structure and DNA methylation is significantly reduced by the time germ cells have entered the genital ridge. During subsequent stages of development, when oocytes and spermatozoa are formed, de novo methylation of DNA takes place and imprints are repositioned in a sex-specific manner and at distinct times. This is the first step of a process referred to as epigenetic reprograming that involves wide demethylation and remethylation. As a result of these processes the genomes contained in each gamete are complementary and not equivalent. It is interesting to note that, due to the epigenetic modifications, oocytes are not totipotent in mammals and a paternal genome is essential to 'rescue' the oocyte, as the maternal genes are imprinted reciprocally to paternal imprints (Fig. 1.14).

The mechanisms that control DNA demethylation in PGCs and the subsequent erasure of imprints have not been well established. Immunofluorescence studies have shown that early PGCs possess unique chromatin marks that distinguish them from the surrounding somatic cells even prior to migration. Experiments carried out with germ cells cultured in vitro without gonadal somatic cells, indicated PGC ability to undergo allele specific DNA methylation suggesting their capacity for

Table 1.3 List of maternal or paternal imprinted genes and their chromosomal location

Gene	Chromosome	Expressed allele
TP73	1	Maternal
DIRAS3	1	Paternal
NAP1L5	4	Paternal
PLAGL1	6	Paternal
HYMAI	6	Paternal
SLC22A2*	6	Maternal
SLC22A3*	6	Maternal
COPG2IT1	7	Paternal
DDC	7	Isoform Dependent
GRB10	7	Isoform Dependent
TFPI2	7	Maternal
SGCE	7	Paternal
PEG10	7	Paternal
PPP1R9A	7	Maternal
DLX5	7	Maternal
CPA4	7	Maternal
MEST	7	Paternal
MESTIT1	7	Paternal
KLF14	7	Maternal
DLGAP2	8	Paternal
KCNK9	8	Maternal
ABCA1	9	
INPP5F V2	10	Paternal
KCNQ10T1	---	Paternal
H19	11	Maternal
IGF2	11	Paternal
IGF2AS	11	Paternal
INS	11	Paternal
KCNQ1	11	Maternal
KCNQ1DN	11	Maternal
CDKN1C	11	Maternal
SLC22A18	11	Maternal
PHLDA2	11	Maternal
OSBPL5	11	Maternal
WT1-Alt trans	11	Paternal
RBP5	12	Maternal
MEG3	14	Maternal
DLK1	14	Paternal
PWCR1	15	Paternal
NDN	15	Paternal
SNURF	15	Paternal
SNORD107	15	Paternal
SNORD64	15	Paternal
SNORD108	15	Paternal
SNORD109B	15	Paternal
MKRN3	15	Paternal
MAGEL2	15	Paternal
SNRPN	15	Paternal
SNORD109A	15	Paternal
SNORD115@	15	Paternal

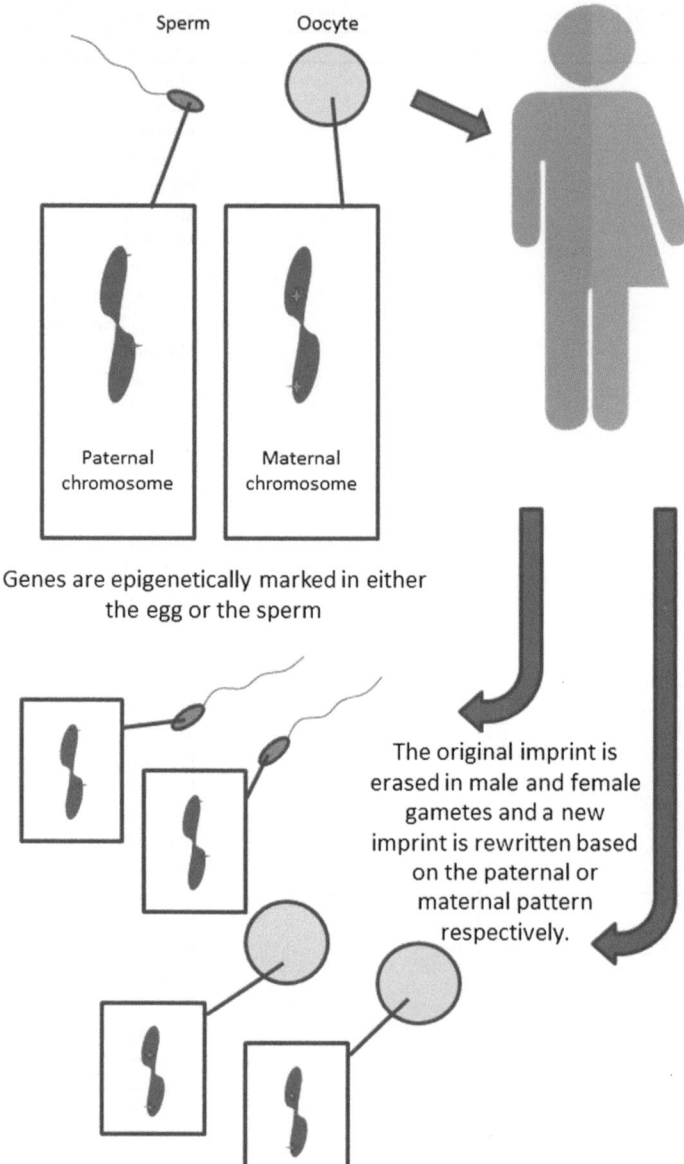

Fig. 1.13 Representative figure of maternal and paternal genomic imprinting in mammalian species

epigenetic reprograming is intrinsic and independent of the genital ridge environment. On the other hand, recent evidence demonstrates that the somatic follicle (Female) and Sertoli cells (Male) directly contribute to imprint establishment.

Because of imprinting, normal development requires the presence of both the maternal and paternal genomes. Indeed, mouse obtained from two maternal or

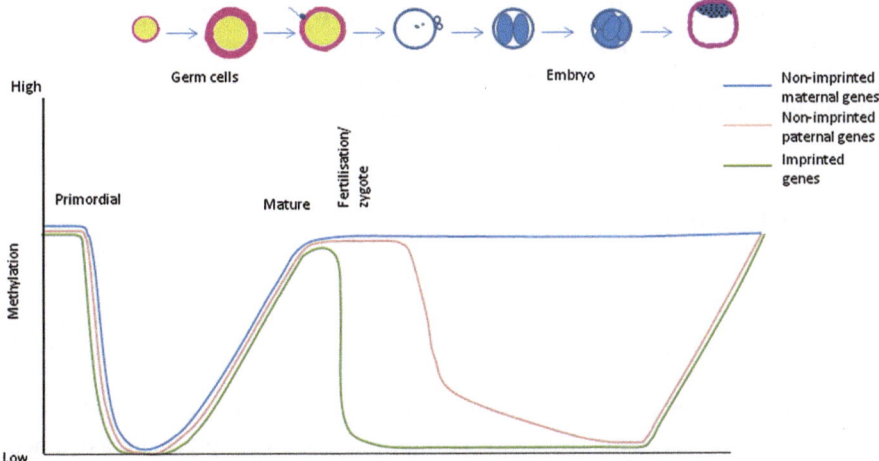

Fig. 1.14 Epigenetic modification pattern. The first step of epigenetic reprograming appears during the primordial germ cell development. The second step of demethylation and remethylation occurs after fertilization

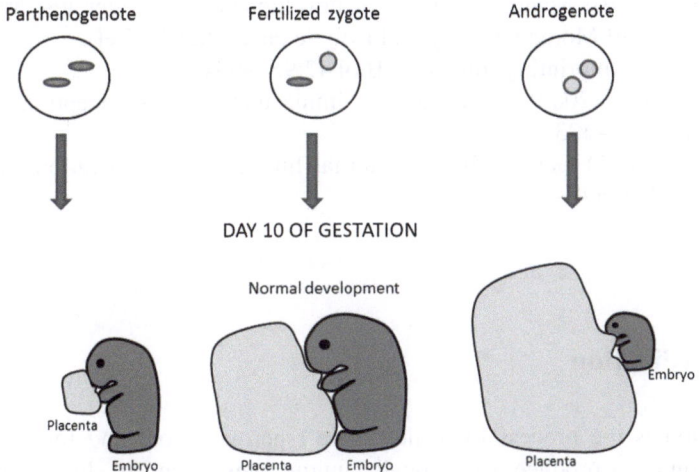

Fig. 1.15 Phenotype scheme of parthenogenetic, fertilized, and androgenetic murine embryos after 10 days of development. Parthenotes display a good embryonic development and a poor placental growth, compared with fertilized embryos. In contrast, the presence of two paternal genomes (androgenote) causes a limited growth of the embryo proper but a better placental growth

egg genomes (parthenogenotes) or with two paternal or sperm genomes (androgenotes) die before or around the implantation stage. It was also observed that the very few parthenogenetic embryos that developed to post-implantation stages, displayed a better embryonic development but a poor placental growth. In contrast, androgenotes had a well-developed placenta but a fairly limited growth of the embryo proper. It is at present accepted that no naturally occurring cases of parthenogenesis exist in mammalian species due to imprinting that would prevent a correct implantation (see chapter of parthenogenesis) (Fig. 1.15).

Further Reading

Anthony JF Griffiths, Jeffrey H Miller, David T Suzuki, Richard C Lewontin, William M Gelbart (2000) An introduction to genetic analysis, 7th edn. W. H. Freeman, New York

Bartolomei MS (1994) The search for imprinted genes. Nat Genet 6:220–221

Ferguson-Smith AC, Surani MA (2001) Imprinting and the epigenetic asymmetry between parental genomes. Science 293(5532):1086–9

Hyttel P, Sinowatz P, Vejlsted M, Betteridge K (2010) Essentials of domestic animal embryology. Saunders Elsevier, Philadelphia

Koniukhov BV, Platonov ES (2001) Genomic imprinting in mammals. Genetika 37:5–17

Marlow FL (2010) Maternal control of development in vertebrates. My mother made me do it! Morgan & Claypool Life Sciences, San Rafael

Solter D (1998) Imprinting. Int J Dev Biol 42:951–954

Surani MA (2002) Genetics: immaculate misconception. Nature 416(6880):491–493

Surani MA (2007) Germ cells: the eternal link between generations. C R Biol 330(6–7):474–478

1.5 Fertilization

Fertilization is the process by which male (spermatozoon) and female (oocyte) gametes unite to produce a genetically unique new organism. In mammals, this event occurs in the ampullary region of the oviduct. During copulation ejaculate, containing a large number of spermatozoa, is deposited into the cranial vagina (sheep, cat, rabbit, dog, cow, and primates) or directly into the cervix (pig and horse). In a few minutes, sperm cells move into the ampulla thanks to contractions of the tunica muscularis of the female reproductive tract and to estrogen-induced changes in cervical mucus, which result in the formation of channels that facilitate movement of sperm.

However, at this point, ejaculated spermatozoa are not able to fertilize the oocyte and require specific modifications which occur during capacitation, a process that may take place in different sites of the female reproductive tract depending on the species.

Capacitation triggered by bicarbonate ions (HCO_3^-) that enter the sperm and directly activate a soluble adenylyl cyclase enzyme in the cytosol has been correlated with several changes:

1. removal of the seminal plasma proteins and the glycoprotein coat—which interfere with fertilization—from the plasma membrane of sperm cells;
2. flagellar mobility alterations that are necessary for penetrating the zona pellucida;
3. activation of pathways necessary to induce the acrosome reaction;
4. increase of intercellular calcium and hyperpolarization of the membrane potential;
5. various metabolic changes.

After these modifications, the spermatozoon is ready to interact with the oocyte, it migrates through the layer of follicle cells and binds its anterior head receptor with a glycoprotein of the zona pellucida, known as the zona pellucida C (ZPC), the mammalian protein equivalent to mouse ZP3. This interaction induces the acrosome reaction during which hydrolytic enzymes, such as hyaluronidase and acrosin, are released from the acrosome cap of the sperm cell and tail movements become flagellar beat and drive sperm in a forward direction. The spermatozoon penetrates the zona pellucida and adheres to the oocyte plasma membrane thanks to a transmembrane protein, called fertilin. The latter binds to female gamete integrins and to the microvilli of the egg surface that rapidly elongate and cluster around it, ensuring a stable association. Therefore, the spermatozoon enters the egg, through a process known as syngamy, causing oocyte activation. This leads to the establishment of a polyspermic fertilization block, the resumption of meiosis, and to the initiation of the embryonic developmental program.

The activation mechanism is characterized by a fundamental increase, also described as a "wave", of cytosolic calcium (Ca_2^+) concentration that may occur within several seconds or a few minutes following gamete fusion, depending on the species. This event seems to be one of the major causes of oocyte release from its meiotic arrest, together with the prevention of polyspermic fertilization. This last event is also made possible by two processes that hinder further sperm entry:

1. the cortical reaction, an exocytosis of cortical granules that contain proteases, peroxidase, mucopolysaccharides, plasminogen activator, and acid phosphatases;
2. the rapid depolarization of the egg plasma membrane.

Once fertilized, the oocyte is called a zygote. Its activation causes the resumption and accomplishment of meiotic division. The second polar body is extruded and the definitive oocyte shows a female pronucleus with a haploid set of chromosomes (Fig. 1.16).

It is also essential to highlight that, in non-rodent mammalian species, the sperm contributes, not only paternal DNA to the zygote, but also to the centrioles.

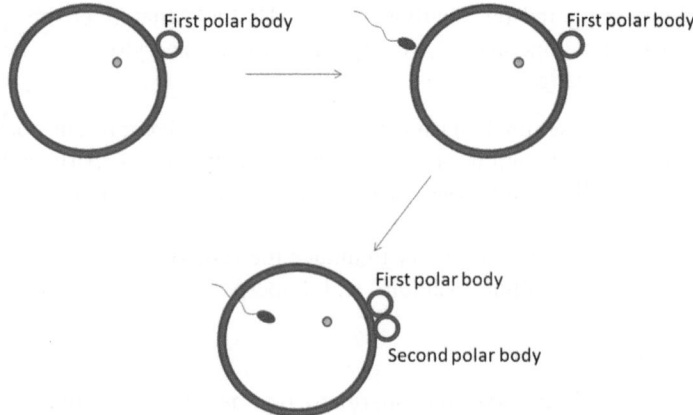

Fig. 1.16 Fertilization and completion of meiosis. The ovulated oocyte is arrested at metaphase II and, only following sperm entry, it resumes meiotic process. Following fertilization, the second polar body is extruded and the oocyte shows a female pronucleus with a haploid set of chromosomes

This organelle, lost by the oocyte during maturation, is fundamental for the organization and assembly of the first mitotic spindle in the zygote.

The sperm nucleus undergoes transformations, it approaches the female pronucleus toward the center of the egg, remaining distinct until after each pronucleus membrane has broken down to form a diploid zygote. Both pronuclei DNA decondense and rapidly replicate to prepare the zygote for its first mitotic division. The embryo undergoes a series of mitotic divisions, also known as embryonic cleavages, progressing through 2-cell, 4-cell, 8-cell, 16-cell, morula, and blastocyst stages.

Further Reading

Alberts B, Johnson A, Lewis J, Raff M, Roberts K, Walter P (2002) Mitosis. Molecular biology of the cell. 4th edn. Garland Science, New York
Hyttel P, Sinowatz P, Vejlsted M, Betteridge K (2010) Essentials of domestic animal embryology. Saunders Elsevier, Philadelphia
Gilbert SF (2000) Developmental biology, 6th edn. Sinauer Associates, Sunderland

1.6 Oocyte Activation

Parthenogenesis literally means "virgin birth". It is a form of asexual reproduction in which an oocyte can develop without the intervention of the male counterpart (without fertilization). This process may naturally occur in many plants and in

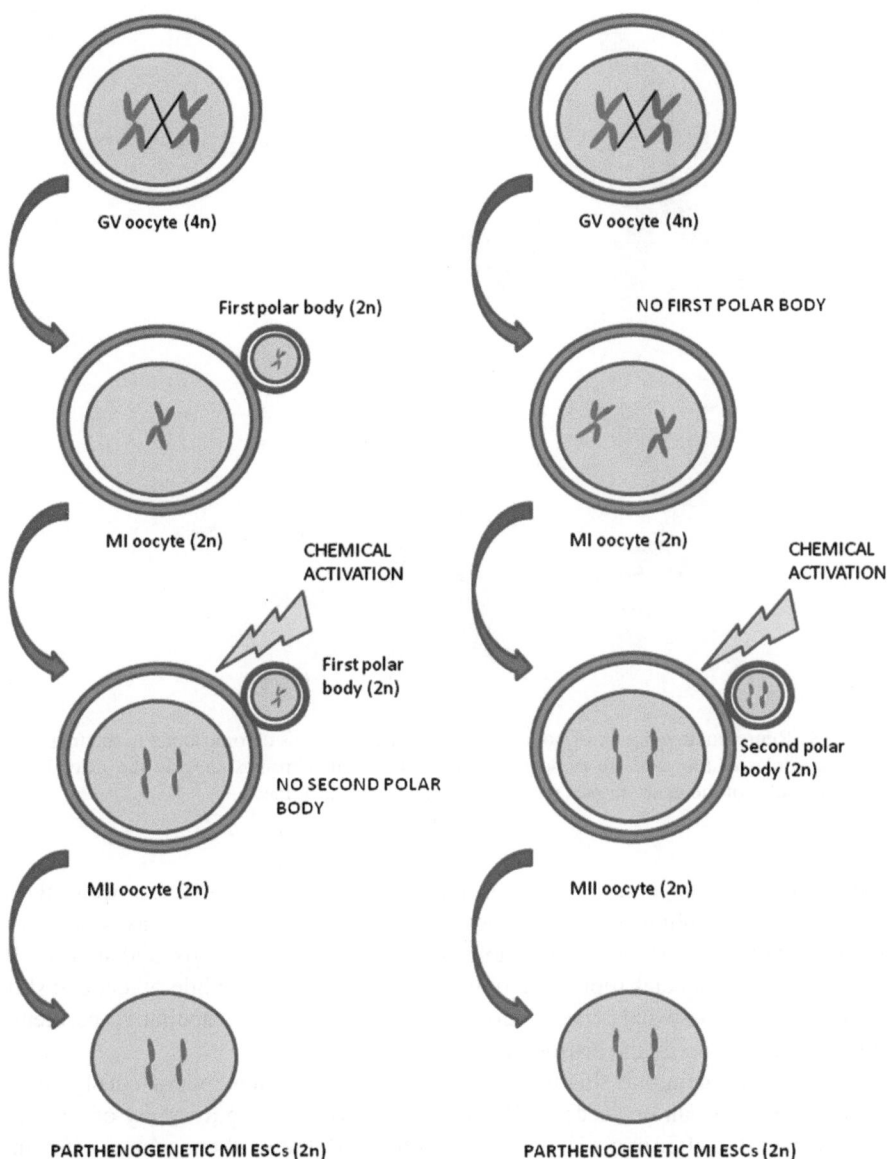

Fig. 1.17 Genetic makeup of parthenotes obtained using two different approaches. When the extrusion of the second polar body is inhibited, parthenotes display high homozygosity, since the diploid status is acquired from the segregation of sister chromatids and the degree of heterozygosity depends only on the extent of crossing over taking place during the prophase of the first meiotic division. By contrast, extrusion of homologous chromosomes does not take place when activation is carried out blocking the extrusion of the first polar body. In this case, a tetraploid oocyte is initially obtained and the diploid status is reached after the extrusion of the second polar body. Only at this stage does the segregation of sister chromatids occur leading to a parthenote that shows a chromosome set identical to that observed in the oocyte

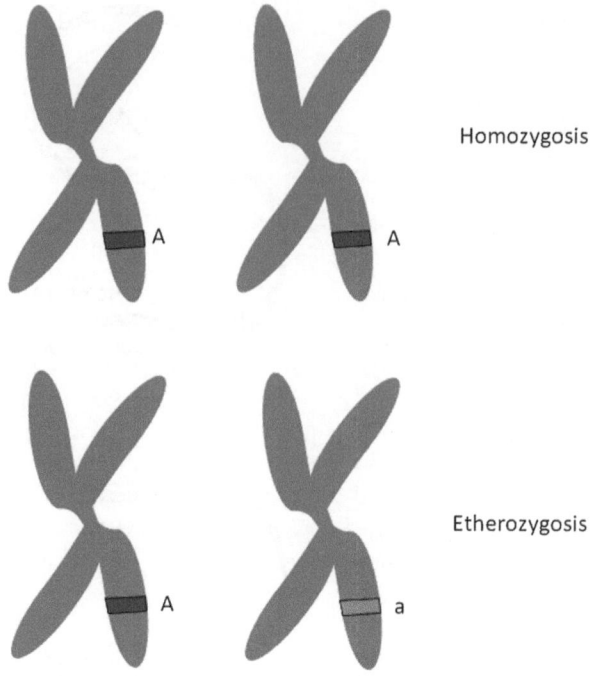

Homozygosis

Etherozygosis

Fig. 1.18 Representative figure of homozygous and heterozygous chromosome. In the first case, identical alleles of the gene are present on both homologous chromosomes. In the second, two different alleles of the gene are present on homologous chromosomes

about 2000 animal species such as invertebrates (including nematodes, water fleas, some scorpions, aphids, some bees, some Phasmida, and parasitic wasps) and vertebrates (including fish, ants, flies, honeybees, amphibians, lizards, and snakes). In particular, some species reproduce only by parthenogenesis, while others can shift between standard sexual reproduction and parthenogenesis (facultative parthenogenesis) depending on the season or the lack of males.

However, in mammals this form of reproduction does not spontaneously occur, but their oocytes can be successfully activated in vitro, using a variety of stimulations. Generally, after meiosis, the oocyte is haploid, but parthenogenetic offspring show a diploid karyotype. This result may be obtained by two different methods (Fig. 1.17) and the choice between them has important consequences on the genetic makeup of the parthenote. An important aspect related to parthenogenesis, in fact, is the zygosity. This refers to the similarity of a single locus on the DNA and is used to describe the genotype of a diploid organism. If both alleles are the same, the organism is homozygous for the trait, while if alleles are different, the organism is heterozygous (Fig. 1.18).

More in detail, parthenogenesis can be induced by inhibition of the extrusion of the first polar body or by preventing the extrusion of the second polar

Fig. 1.19 Comparison
between IVF and
parthenogenetic activation

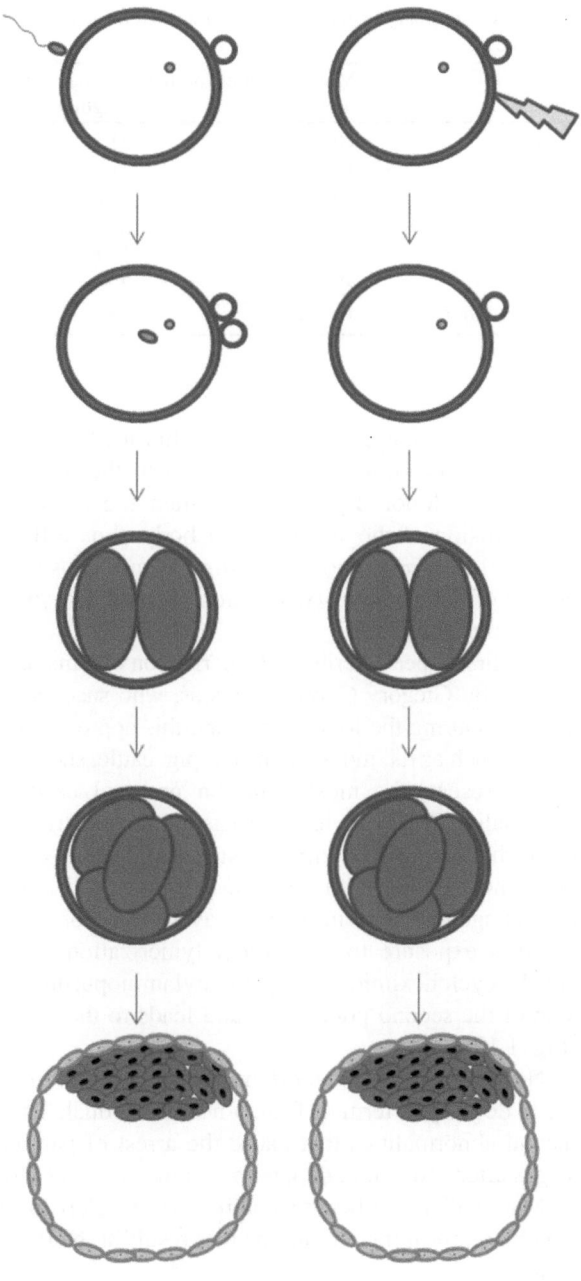

body. In the first case, parthenogenesis is induced in a tetraploid oocyte, whereby the homologous chromosomes do not split yet, and the diploid status is reached after the extrusion of the second polar body, when the segregation

Table 1.4 Summary of maximum development of mammalian parthenotes and length of normal pregnancy

Species	Maximum development (days)	Pregnancy length (days)	References
Mouse	10	21	Surani et al. 1986
Rabbit	10–11	31	Ozil 1990
Pig	29	114	Kure-bayashi et al. 2000
Sheep	25	150	Loi et al. 1998
Bovine	48	280	Fukui et al. 1992
Marmoset monkey	10–12	144	Marshall et al. 1998

of sister chromatids takes place. This leads to the production of a parthenote that has a chromosome set identical to the one observed in the oocyte and in the oocyte donor. By contrast, when the diploid status is obtained blocking the extrusion of the second polar body, thus following the segregation of sister chromatids, a parthenote with high homozygosity is obtained. Here, in fact, the degree of heterozygosity depends only on crossing over that takes place during the prophase I.

The first paper describing the derivation of a mammalian parthenote was published in 1936 by Gregory Goodwin Pincus, who successfully induced parthenogenesis in a rabbit. During the following years, this approach was applied to other mammalian species such as rat, mouse, hamster, pig, cattle, sheep, monkey, and human.

At present, the most common protocol used to induce parthenogenesis in mammalian MII oocytes consists of an electrical or chemical (ethanol, strontium chloride, or ionomycin) stimulus, which mimics the intracellular calcium wave induced by sperm entry in normal fertilization. This phenomenon activates parthenogenesis and initiates cleavage divisions. At the same time, the protocol involves exposure to an actin polymerization inhibitor, usually cytochalasin D and B, cycloheximide or 6-dimethylaminopurine (DMAP) that blocks the extrusion of the second polar body and leads to the generation of a diploid parthenote (Fig. 1.19).

Nevertheless, it is important to highlight that mammalian parthenotes are unable to develop to term or form a new individual. This is a consequence of developmental abnormalities that cause the arrest of parthenote development at different stages after activation, depending on the species (Table 1.4).

Mammalian parthenote inability to develop to term may probably be due to genomic imprinting alterations that result in the repression of paternally expressed genes.

Indeed, while mammals have imprinted genetic regions where either the maternal or the paternal chromosome is inactivated in the offspring in order to proceed normally in its development, a parthenote presents only a double set of maternally imprinted genes and lacks paternal ones. This has been indicated and is, at present, considered the main cause for placentation defects and embryo death.

Further Reading

Brevini TA, Gandolfi F (2008) Parthenotes as a source of embryonic stem cells. Cell Prolif 41(Suppl 1):20–30

Ducibella T et al (2002) Egg-to-embryo transition is driven by differential responses to Ca(2 +) oscillation number. Dev Biol 250:280–291

Kubiak J, Paldi A, Weber M, Maro B (1991) Genetically identical parthenogenetic mouse embryos produced by inhibition of the first meiotic cleavage with cytochalasin D. Development 111:763–769

Mann MR et al (2004) Selective loss of imprinting in the placenta following pre-implantation development in culture. Development 131:3727–3735

Rougier N, Werb Z (2001) Minireview: parthenogenesis in mammals. Mol Reprod Dev 59:468–474

Surani MA, Barton SC (1983) Development of gynogenetic eggs in the mouse: implications for parthenogenetic embryos. Science 222:1034–1036

Surani MA, Barton SC, Norris ML (1984) Development of reconstituted mouse eggs suggests imprinting of the genome during gametogenesis. Nature 308:548–550

Wolffe AP, Matzke MA (1999) Epigenetics: regulation through repression. Science 286:481–486

Chapter 2
Early Embryo Development

2.1 Syngamy and Spindle Formation

Following the entry of the spermatozoon, the oocyte undergoes the activation process that stimulates the second polar body extrusion. At this stage the female pronucleus, containing a haploid set of chromosomes, is formed. Its male counterpart, the male pronucleus, increases its volume due to chromatin decondensation. This event involves a reduction of disulphite cross-links, made by glutathione, and a replacement of sperm polyamines with histones from the oocyte. Finally, male and female pronuclei approach each other, thanks to the male pronucleus association with the sperm centrosome. The latter organizes the aster that captures the female pronucleus, drawing it toward the male pronucleus, and moves both of them toward the center of the oocyte. At the time of nuclear envelope dismantling, pronuclei are fused and syngamy takes place. The diploid genome of the zygote is thus created (Fig. 2.1).

To better understand the importance of the role of the sperm centrosome during fertilization, it is necessary to take a step back and carefully consider the gamete maturation process. The centrosome, approximately 1 μm in size, is a non-membrane-bound cytoplasmatic organelle composed by several protein complexes (Fig. 2.2). Its main function is the organization of interphase microtubule arrays and mitotic/meiotic spindles.

Originally, spermatids and primary oocytes display a typical centrosome organization with a pair of centrioles surrounded by pericentriolar material, in common with somatic cells. These structures undergo extensive modification and/or degeneration during gametogenesis. In non-rodent species, spermatozoa preserve centrioles but loose most of the pericentriolar centrosomal proteins, whereas oocytes waste centrioles and retain only a stockpile of centrosomal proteins. This reciprocal reduction of centrosomal constituents makes sperm and oocyte complementary to each other. In this way they become able to form a functional centrosome in the zygote only after fertilization.

In non-rodent species, centrosomes are reduced during spermiogenesis. Their mature spermatozoa have intact proximal centrioles, whereas the distal centrioles

T. A. L. Brevini and G. Pennarossa, *Gametogenesis, Early Embryo Development, and Stem Cell Derivation*, SpringerBriefs in Stem Cells, DOI: 10.1007/978-1-4614-5532-5_2, © The Author(s) 2013

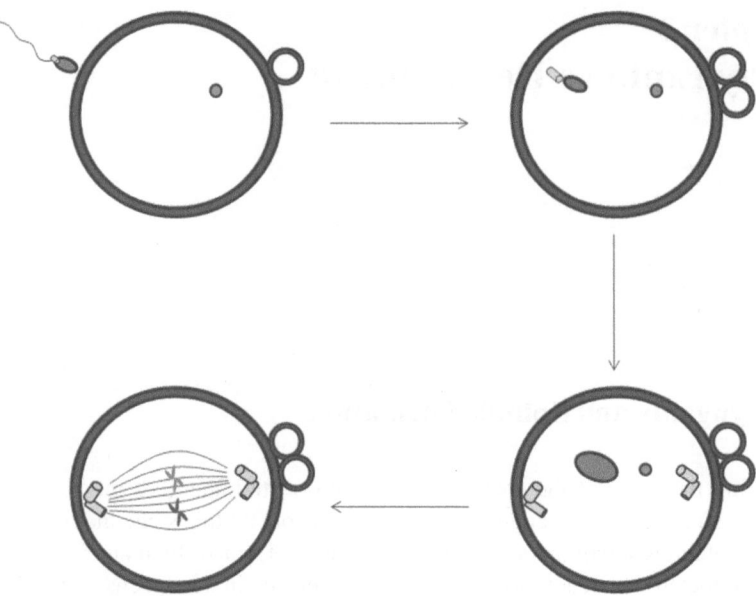

Fig. 2.1 The process of syngamy. After sperm entry, the oocyte extrudes the second polar body. The male pronucleus increases its volume due to chromatin decondensation and approaches the female pronucleus, moving toward the center of the oocyte. At the time of nuclear envelope dismantling, pronuclei are fused and syngamy takes place. The diploid genome distinctive of the zygote is created

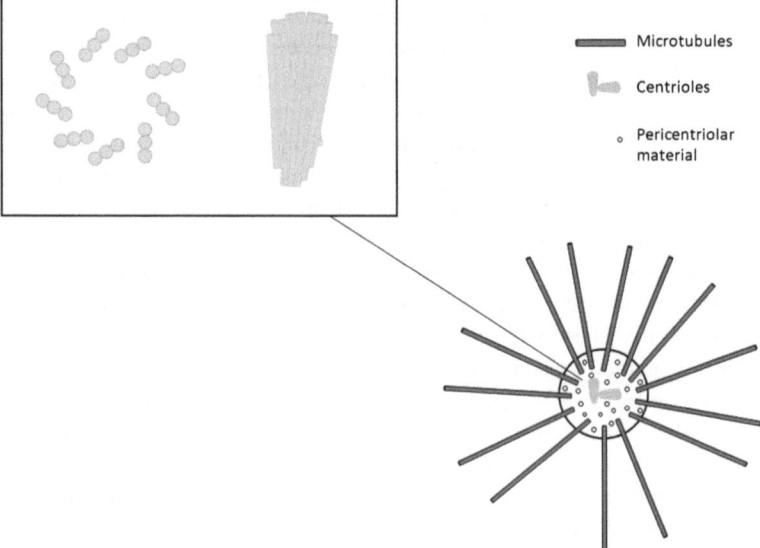

Fig. 2.2 Centrosome structure. The centrosome consists of a pair of orthogonally arranged cylinder-shaped centrioles (*yellow*) surrounded by an amorphous matrix of electron dense proteins referred to as pericentriolar material (*white circle*). Centrioles show the classic 9 + 0 pattern of nine triplet microtubules and no central pair of microtubules and contain centrin, cenexin, and tektin. The pericentriolar material contains a complex of proteins responsible for microtubule (*green*) nucleation

Fig. 2.3 Centrosome reduction during spermiogenesis. Male germ cells possess intact centrosomes containing centrioles and centrosomal proteins until the round spermatid stage. During spermiogenesis, centrosomal proteins are disjuncted from the centrioles and discarded with the residual bodies. Eventually, rodent and snail mature spermatozoa completely loose both centrioles, whereas non-rodent mammalian spermatozoa retain an intact proximal centriole but degenerate the distal one

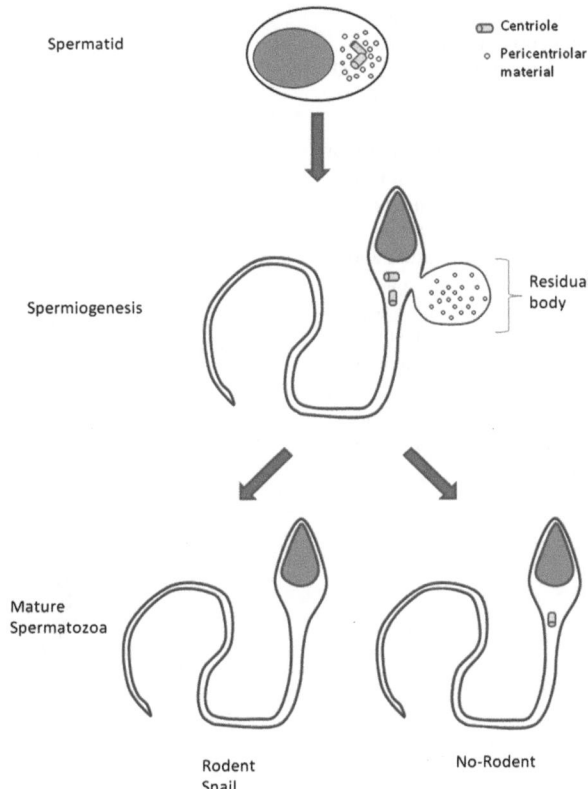

are mostly disorganized or highly degenerated together with γ-tubulin and centrosomal proteins (Fig. 2.3).

By contrast, in rodents and snails, microtubules lose their nucleating function and centrosomal proteins are discarded during spermiation. The distal centriole degenerates during the testicular stage of spermiogenesis and the proximal centriole is lost during the epididymal stage (Fig. 2.3).

In female gametes, centrioles disappear during early oogenesis. Oogonia and fetal oocytes display normal centrioles until the pachytene stage, whereas these organelles are absent in mature oocytes (Fig. 2.4). This degenerative process has been demonstrated in many non-rodent species, including human, rhesus monkeys, rabbits, sheep, cow, and pig, as well as in lower species, such as sea urchins and Xenopus. Similarly, in rodents, centrioles completely degenerate during oogenesis and, at the time of germinal vesicle breakdown, multiple foci composed of perinuclear material, which gradually coalesce to form the poles of metaphase spindles, appear.

At fertilization, sperm and egg equally contribute haploid genomes as well as the relative centrosome components. Since in non-rodent mammals the MII oocyte centrosome is degraded and centrioles are absent, as previously described, early embryo development requires maternal and paternal

Fig. 2.4 Centrosome reduction during oogenesis. Oogonia possesses standard centrosomes containing centrioles and centrosomal proteins. Mammalian primary oocytes lose both centrioles completely, resulting in acentriolar and anastral poles during meiotic I and II divisions. The pericentriolar centrosomal proteins are dispersed in the oocyte cytoplasm during the non-dividing stages or distributed as concentric poles of the barrel-shaped spindles during dividing stages

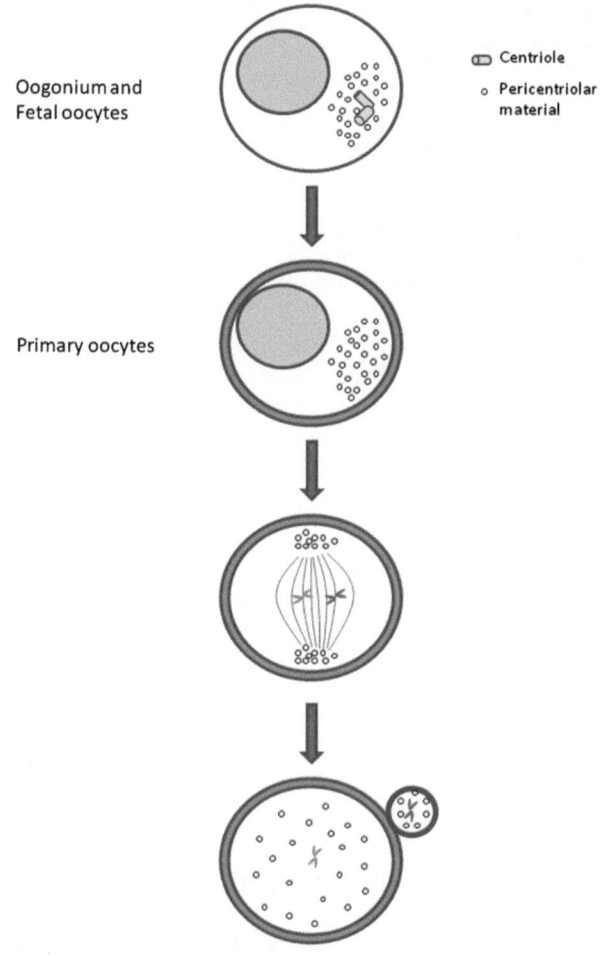

Oogonium and Fetal oocytes

Primary oocytes

⊞ Centriole

∘ Pericentriolar material

contribution and, in particular, requires their elements to restore a normal and functional centrosome. Indeed, the sperm tail and its proximal centriole are introduced into the ooplasm together with the sperm head (Fig. 2.5). Paternal pronucleus decondenses in the ooplasm and the proximal centriole remains intact forming the aster, while most of the other sperm cytoplasmic structures including mitochondria, microtubule, and fibers are eliminated. The sperm aster enlarges and moves within cytoplasm, ensuring male and female pronuclei apposition. Although centrioles are paternally inherited, the formation of a single mitotic metaphase plate, with a bipolar spindle, requires the interaction between the sperm centriole and the with maternal pericentriolar proteins. In agreement with this, several studies have demonstrated that human sperm centrioles duplicate during the pronuclear stage, and at syngamy, centrioles are located at

Fig. 2.5 Centrosome inheritance in mammalian species. In non-rodent mammalian species (*left panel*), sperm contains a proximal centriole before fertilization, while MII oocytes show a meiotic spindle with acentriolar centrosomes. The proximal centriole is introduced by the sperm and replicates forming the aster. At syngamy, it relocates to opposite poles to form the centers of the mitotic spindle poles and drives the first embryo mitotic duplication. By contrast, in rodents and in Drosophila, both sperm and oocyte lose their centrioles and only the centrosomal material is maternally inherited. This material remains dispersed in oocytes and after fertilization zygotes show centrosomal proteins aggregate forming aster-like structures

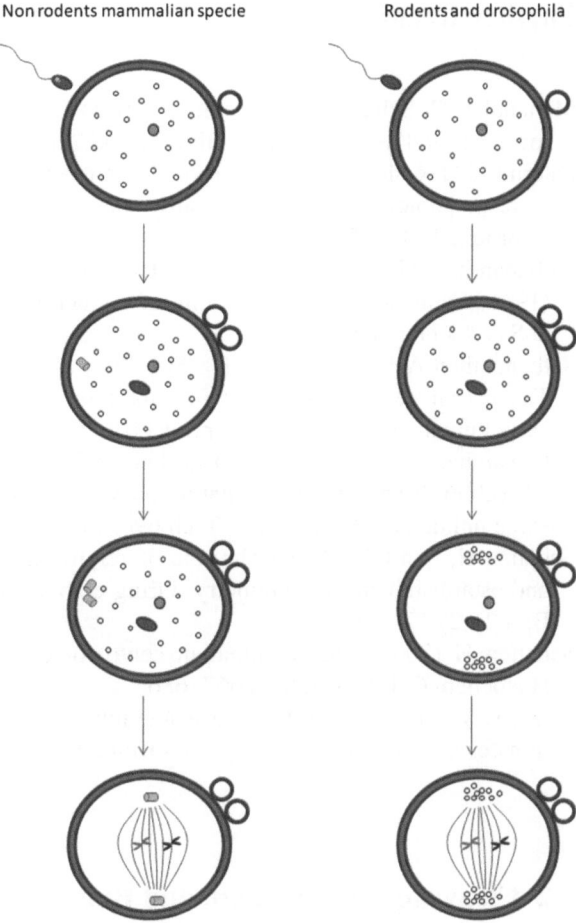

Non rodents mammalian specie   Rodents and drosophila

opposite poles of the first mitotic spindle, together with a surrounding pericentriolar material of the oocyte that nucleates microtubules. The same process was also described in sea urchin, rabbit, cattle, and sheep, whereby following insemination, microscope observations revealed one bipolar spindle displaying two centrioles at the opposite pole (first embryo cleavage). Together, these observations suggest that zygote centrosomes represent the origin of embryonic, fetal, and adult somatic cell centrosomes.

Conversely, in mouse and Drosophila, spermatozoa do not show the presence of centrioles and the centrosomes are maternally inherited, remaining dispersed in oocytes. Indeed, after sperm entry, typical centriole structures are absent in murine zygote, and early embryos show centrosomal proteins that aggregate in clusters forming small aster-like arrays of microtubules. Only at the blastocyst stage are centrosomes with double centrioles formed de novo.

Further Reading

Manandhar G, Schatten H, Sutovsky P (2005) Centrosome reduction during game-
    togenesis and its significance. Biol Reprod 72(1):2–13
Palermo GD, Colombero LT, Rosenwaks Z (1997) The human sperm centrosome
    is responsible for normal syngamy and early embryonic development. Rev
    Reprod 2(1):19–27
Sathananthan AH, Kola I, Osborne J, Trounson A, Ng SC, Bongso A, Ratnam SS
    (1991) Centrioles in the beginning of human development. Proc Natl Acad Sci
    U S A 88(11):4806–4810
Sathananthan AH, Ratnam SS, Ng SC, Tarín JJ, Gianaroli L, Trounson A (1996)
    The sperm centriolecentriole: its inheritance, replication and perpetuation in
    early human embryos. Hum Reprod 11(2):345–356
Sathananthan AH, Selvaraj K, Girijashankar ML, Ganesh V, Selvaraj P, Trounson
    AO (2006) From oogonia to mature oocytes: inactivation of the maternal centro-
    some in humans. Microsc Res Tech 69(6):396–407
Schatten H, Sun QY (2010) The role of centrosomes in fertilization, cell division
    and establishment of asymmetry during embryo development. Semin Cell Dev
    Biol 21(2):174–184
Schatten H (2008) The mammalian centrosome and its functional significance.
    Histochem Cell Biol 129(6):667–686
Sun QY, Schatten H (2007) Centrosome inheritance after fertilization and nuclear
    transfer in mammals. Adv Exp Med Biol 591:58–71

2.2 Cleavage, Compaction, and Blastulation

The first cleavage after syngamy is usually completed within 24 h after ovulation. After the first mitotic event, the zygote is divided into two cells, known as blastomeres that contain a full copy of the new embryo genome. Subsequently, a series of successive mitotic divisions take place and the embryo remains surrounded by the zona pellucida for several days. During this developmental phase, the embryo maintains the same total volume since the original cytoplasm is split among the newly formed blastomeres that become smaller and smaller with each division (Fig. 2.6).

Furthermore, transcripts and proteins stored in the oocyte during its maturation are steadily used up and degraded starting from fertilization, and embryonic genome activation with new RNAs produced under the direct and exclusive control of the embryo transcriptional machinery is required to continue the development.

Embryonic cleavages occur when the embryo is transported along the maternal oviduct and enters the uterus to implant. Species-specific timing characterizes this event as described in Fig. 2.7.

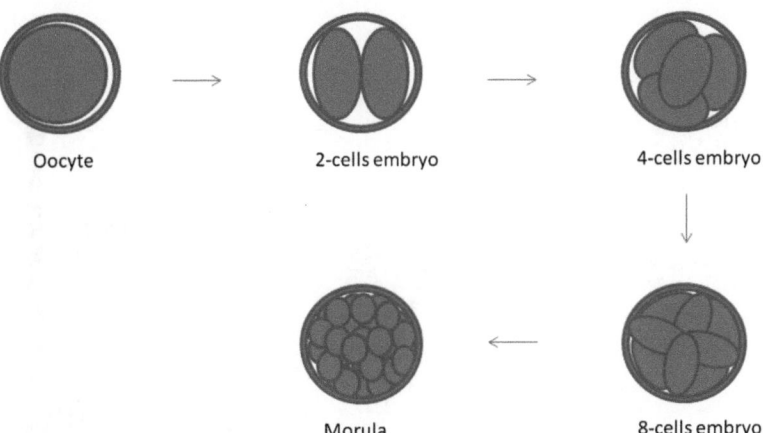

Oocyte   2-cells embryo   4-cells embryo

Morula   8-cells embryo

Fig. 2.6 Embryo cleavage. After fertilization the newly formed zygote remains surrounded by the zona pellucida and mitotically divides. The embryo maintains the same total volume and the original cytoplasm is split among newly formed blastomeres that become smaller and smaller

The following phase is characterized by the first cell differentiation regulated by the activation of specific transcription factors, namely CDX2 and EOMES (Fig. 2.8). In particular, morula cells, that are identical to each other, start to change and the outer cells differentiate in epithelium constituting the trophectoderm or trophoblast. This event is known as compaction and gives the embryo a smoother surface. Trophectoderm cells attach with neighboring cells and form tight junctions and desmosomes. These specialized intercellular structures contribute to intercellular sealing and tissue integrity, critical for vectorial transport and blastocoel cavity formation.

The step that follows morula compaction is known as blastulation phase, during which the embryo transforms into a blastocyst. The trophoblast cells secrete a fluid into the central cavity—the blastocyst cavity or blastocoel—lining the cavity. The inner cells move toward one pole, forming the inner cell mass (ICM). These cells maintain their pluripotent state and will form the embryo proper, while the trophectoderm cells will give rise to the embryonic placenta. The blastocoel continues to gradually increase its volume and the blastocyst expands. This process is made possible thanks to the activity of a sodium pump located in the cell membranes of the trophoblast cells that osmotically draws water into the central cavity. The zona pellucida is still present, but its components will be soon lysed by a blastocyst-secreted protease, known as strypsin, and by proteolytic enzymes produced by the endometrium. The process that leads to zona pellucida degradation is described as hatching and enable the trophoblast cells to directly bind the uterine cavity (Fig. 2.9).

During blastulation, ICM cells further differentiate into two cells population as described in Fig. 2.10:

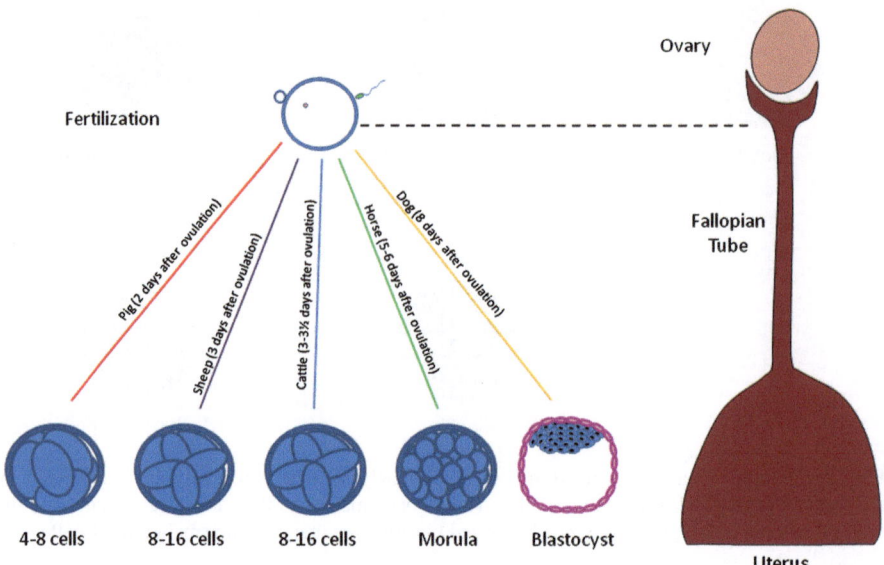

Fig. 2.7 Times and related stages of embryo passage from the oviduct into the uterus. The oocyte is fertilized in the ampulla of the uterine tube and the newly formed embryo mitotically divides during its transport along the oviduct. The time of entry in the uterus and the embryo stage are species-specific

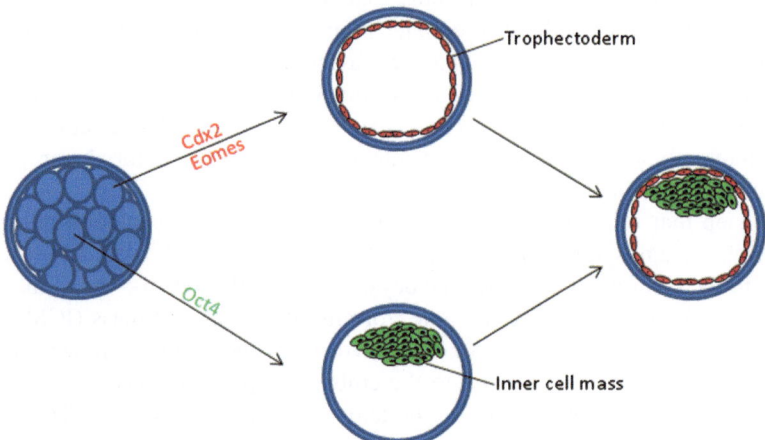

Fig. 2.8 Scheme of initial differentiation. After morula compaction, blastomeres begin to differentiate. Some cells activate the transcription of *CDX2* and *EOMES*, which induce trophectoderm cell formation. Others continue to express OCT4 that maintains the pluripotent status of inner cell mass cells

1. Hypoblast cells: small epithelial cuboidal cells closer to the blastocoel that will form the inner epithelium of the yolk sac. They need the activation of GATA-binding factor 6 (GATA-6), a fundamental transcription factor driving the formation of the primitive endoderm.

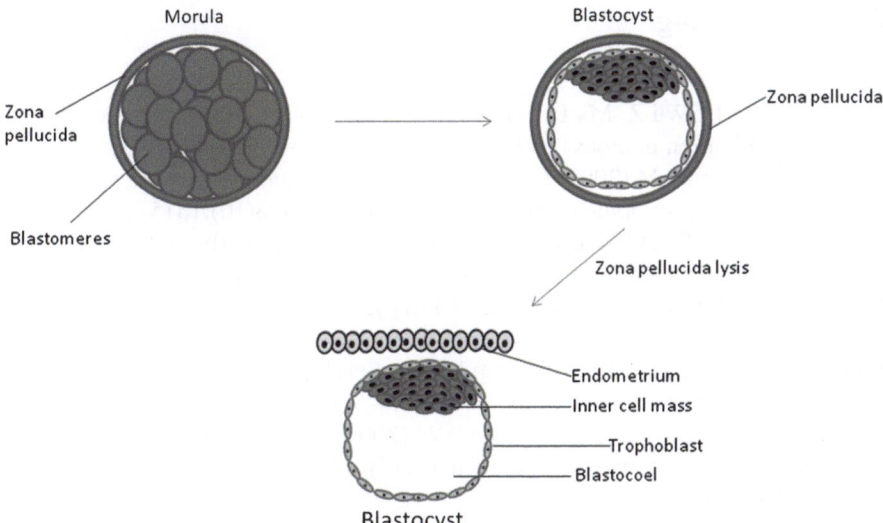

Fig. 2.9 Early embryo development and hatching. Trophoblast cells secrete a fluid into the central cavity, forming the blastocoel. The pluripotent inner cells move toward one pole, forming the inner cell mass. Finally, the zona pellucida is lysed by proteolytic enzymes (hatching), enabling the trophoblast cells to directly bind the uterine cavity (endometrium)

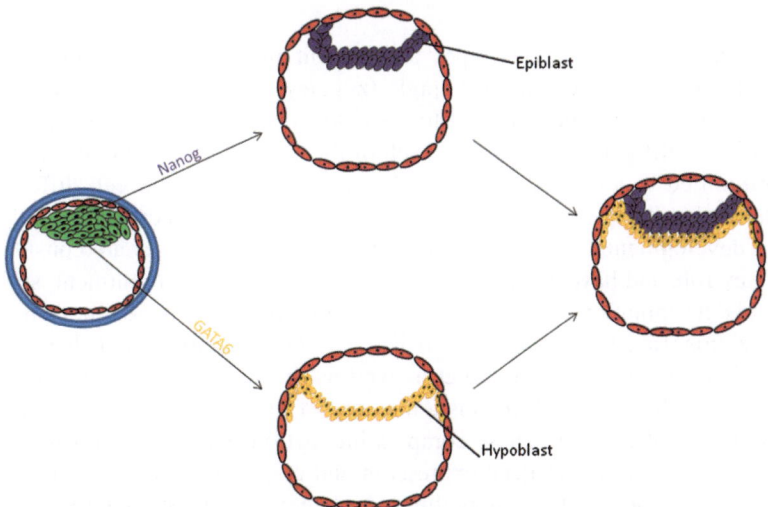

Fig. 2.10 ICM cell differentiation. During blastulation ICM pluripotent cells differentiate into epiblast and hypoblast cells, through the activation of *Nanog* and *GATA6* genes, respectively

2.   Epiblast cells: pluripotent cells that express NANOG and form the primary ectoderm. They can also differentiate into all cells of the three layers during the gastrulation process, giving rise to the embryo proper and all the different tissue types that can be found in an organism.

Further Reading

Chen L, Wang D, Wu Z, Ma L, Daley GQ (2010) Molecular basis of the first cell fate determination in mouse embryogenesis. Cell Res 20(9):982–993

Roper S, Hemberger M (2009) Defining pathways that enforce cell lineage specification in early development and stem cells. Cell Cycle 8(10):1515–1525

Rossant J, Tam PP (2009) Blastocyst lineage formation, early embryonic asymmetries and axis patterning in the mouse. Development 136(5):701–713

Gilbert SF (2000) Developmental Biology, 6th edn. Sinauer Associates, Sunderland

Senner CE, Hemberger M (2010) Regulation of early trophoblast differentiation— lessons from the mouse. Placenta 31(11):944–950

Zernicka-Goetz M, Morris SA, Bruce AW (2009) Making a firm decision: multifaceted regulation of cell fate in the early mouse embryo. Nat Rev Genet 10(7):467–477

2.3 Cell Commitment and Waddington Model of Epigenetic Restriction: Asymmetric Imprinting

More than 230 different cell types are present in the adult mammalian body. Although they all derive from one single (zygote), they are able to differentiate in a tissue and time specific way and to respond to specific developmental cues. At the end of its differentiation process each of this type of cells is highly specialized and committed to a distinct determined fate. How a particular cell differentiates into its final cell type is still to be fully elucidated and represents a challenging goal of developmental biology. At present, four main processes are considered to play a key role and have been shown to be involved in cell commitment, specification, and determination. These processes are cell proliferation, cell movement, cell specialization, and cell interaction. It has also been demonstrated that each cell constituting an embryo is the target as well as the source of specific cues for its neighboring cells. Each cell retains a memory of its own cell proliferation history and its positional changes. These complex interactions have been shown to be regulated through differential gene expression and epigenetic restrictions that gradually limit cell potency to a more limited phenotype-related expression pattern. These concepts have been nicely depicted by Waddington in his very famous epigenetic landscape where a ball represents a cell of an embryo committing to a certain cell fate by rolling from a non-committed, pluripotent condition down a hill marked by slopes and valleys. In Waddington's metaphor, the hill represents the many different and complex process of the cell differentiation process. All those slopes and valleys eventually address the ball along a progressively more restricted potency pathway, toward a favored position at the bottom of the hill, where the

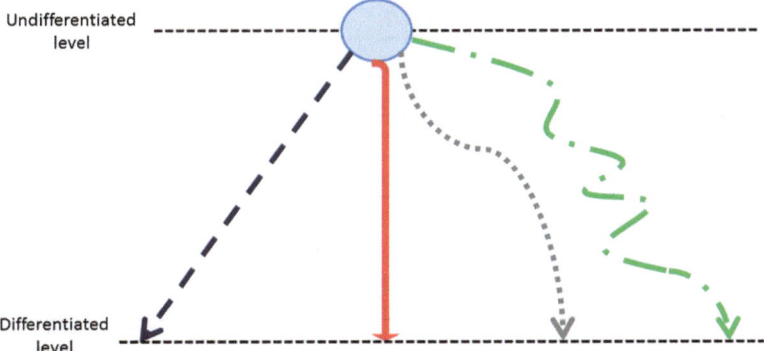

Fig. 2.11 The differentiation slope. According to the original model suggested by Waddington, different developmental canalizations (depicted in the figure as *blue*, *red*, *gray*, and *green lines*) along the differentiation slope, allow an organism to develop from the fertilized egg. The entire set of genes expressed by the differentiating organism and their interactions lead to the composition of a 'developmental system' which produces a phenotype. Interestingly, recent studies have shown that differentiated cells of an organism retain a memory of their own cell differentiation history and positional changes. They can thus be forced in an upstream, counter-current direction up the differentiation slope, along different states of increased potency

cell is unipotent and is characterized by a tissue-specific differentiated state. Many recent studies carried out in stem cells have demonstrated that differentiated cells of our organism can be forced in an upstream, counter-current direction up the differentiation hill, along different states of increased potency. However, it is important to remember that overexpression or activation of additional factors is needed in order to allow epigenetic reprograming events and reach a high plasticity state such as that obtained in induced pluripotent stem (iPS) cells (Fig. 2.11).

In the early embryo or blastocyst we can identify a group of cells, called the inner cell mass, the outer cells referred to as the trophoblast and the blastocoel, which is a fluid filled cavity. As already discussed, inner cell mass gives rise to the embryo proper and consists of cells that have the ability to differentiate in every tissue of the body. Because of this ability they are considered to be pluripotent. The inner cell mass soon develops into a bilaminar structure that comprises the epiblast and the hypoblast. Through the process of gastrulation the epiblast gives rise to all three somatic germ layers of the embryo, ectoderm, mesoderm, and endoderm, and allows for the differentiation of the primordial germ cells. The hypoblast, on the other hand, will lead to the formation of the extraembryonic structures (Fig. 2.12).

Cell specification and differentiation are made possible thanks to decisions driven by many complex process that result from cell-intrinsic properties but, at the same time, need inputs deriving from cell-extrinsic signals.

Cell-intrinsic properties control specification process through an asymmetric cleavage that leads to an unequal distribution of cytoplasmic determinants (proteins, mRNA, etc.) (Fig. 2.13). Once asymmetry is created from homogeneity, the daughter cells become very different in content and are determined to distinct fates. Cell-extrinsic specification is supported by cues deriving from interactions between cells

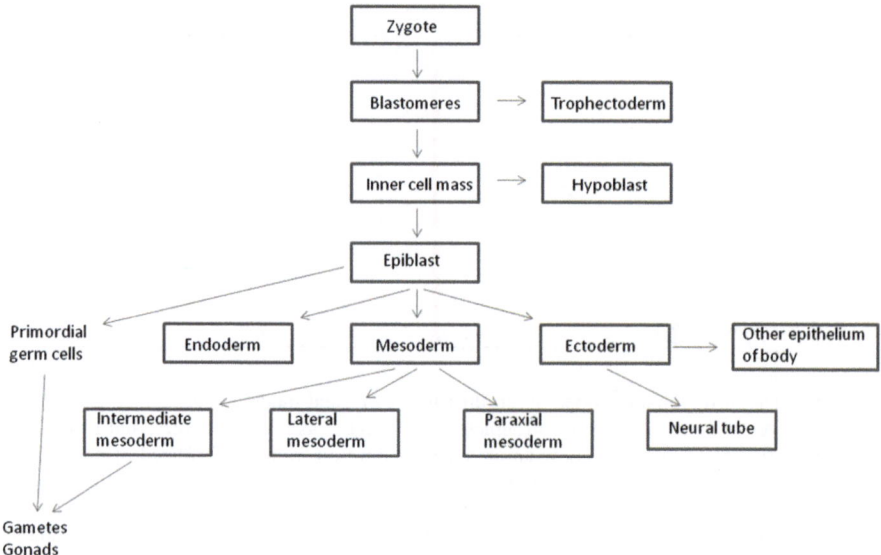

Fig. 2.12 Summary of the first steps of zygote differentiation

Fig. 2.13 The French flag model. Based on this model, the zone of polarizing activity (ZPA), produces a morphogen that diffuses across the nearby areas to generate a spatial gradient

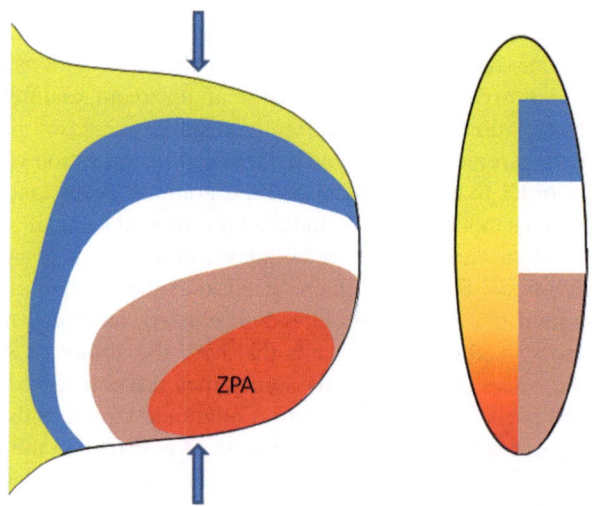

as well as from soluble molecules that can diffuse and carry signals that control cell differentiation decisions in a concentration-dependent fashion, defined as morphogens and normally released in a paracrine fashion. Several major families of morphogens have been characterized and are known to be directly involved in cell induction toward a specific lineage. Many of these molecules are also involved in the formation of specific concentration gradient that drive cells to their correct spatial positioning, thus contributing to patterning and organization of the body axis (see Table 2.1).

Table 2.1 Genes directly involved in cell commitment (morphogens) and their specific functions

Family	Members	Functions
Fibroblast growth factor (FGF)	*FGF1–FGF23*	Play important roles in neurogenesis, axon growth, and differentiation during development of the central nervous system. Promote endothelial cell proliferation and the physical organization of endothelial cells into tube-like structures. They thus promote angiogenesis, the growth of new blood vessels from the pre-existing vasculature. Stimulate repair of injured skin and mucosal tissues by stimulating the proliferation, migration and differentiation of epithelial cells
Hedgehog	Sonic hedgehog (*Shh*), Desert hedgehog (*Dhh*), Indian hedgehog (*Ihh*)	Involved in the developmental pattern formation of various organs, such as the nervous system, muscle, the heart, and the lungs. Hedgehog signaling has also been implicated in the development of several human cancers
Wingless (WNT)	*WNT1–WNT11, WNT16*	Wnt signaling plays a significant role in both the cardiovascular and nervous systems during embryonic cell patterning, proliferation, differentiation, and orientation. Furthermore, modulation of Wnt signaling under specific cellular influences can either promote or prevent the early and late stages of apoptotic cellular injury in neurons, endothelial cells, vascular smooth muscle cells, and cardiomyocytes
Transforming growth factor-β (TGF-β)	*TGF-β, Activin*, bone morphogenetic protein (*BMP*), nodal, glial-derived neurotrophic factor (*GDNF*), inhibin, mullerian inhibitory substance (*MIS*)	The TGF-β family has different important functions related to osteoblast differentiation, neurogenesis, ventral mesoderm specification, angiogenesis, extracellular matrix neogenesis, immunosuppression, apoptosis induction, gonad growth, placenta formation, left–right axis determination

Further Reading

Ferrell JE Jr (2012) Bistability, bifurcations, and Waddington's epigenetic land-scape. Curr Biol 22(11):R458–R466

Hanna JH, Saha K, Jaenisch R (2010) Pluripotency and cellular reprogramming: facts, hypotheses, unresolved issues. Cell 143(4):508–525

Hemberger M, Dean W, Reik W (2009) Epigenetic dynamics of stem cells and cell lineage commitment: digging Waddington''s canal. Nat Rev Mol Cell Biol 10(8):526–537

Gilbert SF (2000) Developmental Biology, 6th edn. Sinauer Associates, Sunderland

Takahashi K (2012) Cellular reprogramming—lowering gravity on Waddington's epigenetic landscape. J Cell Sci

Zeller R, López-Ríos J, Zuniga A (2009a) Vertebrate limb bud develop-ment: moving towards integrative analysis of organogenesis. Nat Rev Genet 10(12):845–858

2.4 Establishment of the Body Axis

Decisions taken in the very early embryo lead to cell commitment and the defi-nition of the different lineages. Although a vast array of information is slowly accumulating, it is not fully understood as yet how multiple types of cells and polarities are generated in an embryo that originates from a single cell. Similarly, we still need to define the origin of the signals that activate cell differentiation and generate asymmetries. At present, early polarity is not thought to depend on asym-mmetric localization of maternal determinants.

In the eight-cell mouse embryo, we have to focus on the process known as compaction that leads to the establishment of an "inside–outside polarity". Compaction lead to the accumulation of specific molecules, such as Par3 (parti-tioning defective 3), Par6 (partitioning defective 6), and atypical PKC (aPKC) to the apical side of the embryo, while localizes Par1 (partitioning defective 1) and E-cadherin to the basolateral side of the cells.

This established inside–outside polarity leads to the activation of a signaling loop, called Hippo in Drosophila and Stk3 (Ser/Thr kinase 3)/Mst in the mouse that inhibits proliferation in the inside cells, or inner cell mass (ICM) and down-regulates expression of trophectoderm (TE)-specific genes such as caudal-related homeobox 2 (*Cdx2*). Two types of cells are therefore present in the blastocyst; TE and ICM cells and it is believed that the decision on one or the other cell fate may depend on epigenetic marks related to differences in histone acetylation and meth-ylation among blastomeres.

The ICM is going to soon segregate two types of cells: the primitive endoderm (PrE) and the epiblast. The PrE is formed by a thin layer of cells located on the

Fig. 2.14 Primitive endoderm formation. Epiblast cells increase the expression of *Fgf4*, that binds *Fgfr2* exhibited by primitive endoderm progenitor cells. This interaction cause the repression of *Nano*g expression and the activation of *Grb2* and *Mapk* genes, resulting in the induction of primitive endoderm specific genes, such as *Gata*6

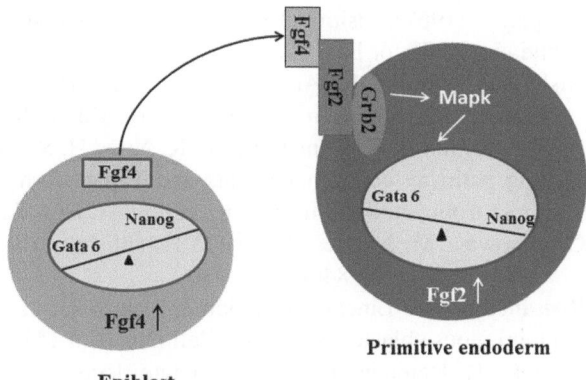

surface of the ICM, facing the blastocoel cavity, and positive for members of the GATA family of transcription factors. The epiblast, on the other hand is inside the ICM and expresses Nanog and Oct3/4. The segregation of the epiblast and PrE lineages in the ICM is known to be regulated by fibroblast growth factor (FGF) and FGF receptor signaling and by the mitogen-activated protein kinase (Mapk) signaling pathway Inhibition of such signaling has been shown to address ICM cells to the epiblast fate (Fig. 2.14).

The formation of the ICM and establishment of the blastocoel cavity creates the embryonic–abembryonic axis, which is at 90° to the margin between the ICM and blastocoel. This axis is considered by several authors as the template for future dorsal–ventral polarity.

The anterior–posterior polarity of the embryo is established in response to signals produced by the visceral endoderm (VE), which is a layer of cells derived from the epiblast that covers the extraembryonic ectoderm and epiblast of the postimplantation embryo. Most VE becomes the endoderm of the extraembryonic yolk sac but a subset of VE cells differentiate into the endoderm of the embryo proper. Molecules produced by the VE act on the nearby epiblast and specify it to future anterior identity. Among them are Lefty 1 and Cer that are Nodal antagonists and specify the fate of the contiguous epiblast portion or anterior visceral endoderm (AVE). By contrast the epiblast region, which is located in a more distant position and is not reached by the AVE derived signals forms the primitive streak on the opposite side of the embryo.

Further Reading

Zeller R, López-Ríos J, Zuniga A (2009b) Vertebrate limb bud development: moving towards integrative analysis of organogenesis. Nat Rev Genet 10(12):845–858

Wolpert L (1969) Positional information and the spatial pattern of cellular differentiation. J Theor Biol 25(1):1–47

Nishioka N, Inoue K, Adachi K, Kiyonari H, Ota M, Ralston A, Yabuta N, Hirahara S, Stephenson RO, Ogonuki N, Makita R, Kurihara H, Morin-Kensicki EM, Nojima H, Rossant J, Nakao K, Niwa H, Sasaki H (2009) The Hippo signaling pathway components Lats and Yap pattern Tead4 activity to distinguish mouse trophectoderm from inner cell mass. Dev Cell 16(3):398–410

Gasperowicz M, Natale DR (2011) Establishing three blastocyst lineages—then what? Biol Reprod 84(4):621–630

Rossant J (2004) Lineage development and polar asymmetries in the peri-implantation mouse blastocyst. Semin Cell Dev Biol 15(5):573–581

Takaoka K, Hamada H (2012) Cell fate decisions and axis determination in the early mouse embryo. Development 139(1):3–14

Chapter 3
Stem cells and Gametogenesis

3.1 Oocyte Competence and Potency

The ability of an individual oocyte to support all the various aspects that allow correct embryo development is referred to as oocyte competence. This concept implies that the female gamete may satisfy the requirements needed for early stage development and, while in the ovary, has accumulated all the nutrients and various molecules that are indispensable for embryogenesis. Indeed, preovulatory oocytes are characterized by a constant increase in transcriptional activity that results in the accumulation of mRNA and proteins that can be promptly used after fertilization and embryogenesis. This closely links the oocyte to the embryo and justifies why the possibility for an embryo to go through preimplantation development and, eventually, to term is largely related to oocyte quality.

It is easy to understand that oocyte competence is rather difficult to define with precision and it is complex to determine. However, it can be summarized as the extent to which an oocyte has completed its growth and has fully saved the entire range of mRNAs and proteins needed until the activation of embryonic transcription. There is an interval between fertilization and the activation of the embryonic genome (when transcriptional activity is turned on and becomes fully functional).

During this time frame, transcription is negligible in the female gamete and the various changes taking place are under the control of maternal RNAs and proteins produced and stored during oogenesis. The length of this period varies according to the species considered. In the mouse, it can occur as early as the late 2-cell stage. In contrast, it happens at the 4-cell stage in pigs, between the 4- and 8-cell stage in human embryos, and between the 9- and 16-cell stage in sheep and bovine embryos (Fig. 3.1). Interestingly, these intervals do not mark the time when embryonic mRNA transcription starts, but rather they have been identified as the cleavage stages that embryos can reach without embryonic transcription (for instance when embryos are exposed to the RNA polymerase II inhibitor α-amanitin).

It is at present largely accepted that oocyte quality and competence depends upon the efficiency of the storing process, and on the appropriate and well-timed

T. A. L. Brevini and G. Pennarossa, *Gametogenesis, Early Embryo Development, and Stem Cell Derivation*, SpringerBriefs in Stem Cells, DOI: 10.1007/978-1-4614-5532-5_3, © The Author(s) 2013

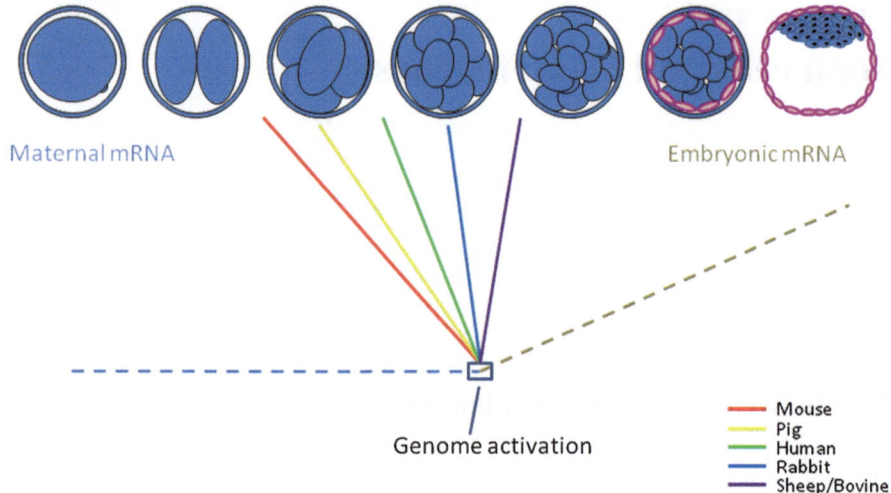

Fig. 3.1 Timing of embryonic genome activation in different species. The preovulatory oocyte accumulates mRNA and proteins that are used immediately after fertilization and during the first stages of embryo development until the activation of embryonic transcription. This activation occurs at different times depending on the species. It starts at the 2-cell stage in mouse, 4-cell stage in pig, at the 4- to 8-cell in human, at 8-cell in rabbit, at 16-cell in sheep and bovine

reactivation of the stored products. However, given the fact that oocytes and early embryos have poor transcriptional activity, availability of encoded proteins is largely under the control of post-transcriptional regulatory mechanisms.

Changes in polyadenylation of specific genes at specific times of development appear like an intriguing mechanism that regulates translational activation/inactivation of maternal transcripts, possibly ensuring the correct concentration of the different molecules at specific times to the developing egg/embryo. It has been shown that regulation of maternal mRNA translation is controlled by changes in the transcript stability that depends on the length of the poly(A) tail stretch at the $3'$ UTR. Oocyte mRNA with short poly(A) tails are translationally inactive. Conversely, mRNAs displaying extended poly(A) tail are stable and translationally active. Changes in polyadenylation of specific genes at specific times of development appear as an intriguing mechanism that regulates translational activation/inactivation of maternal transcripts, possibly ensuring the correct concentration of the different molecules at specific times to the developing egg/embryo. Data from the literature demonstrate a correlation between oocyte developmental competence and polyadenylation patterns. In particular, it has been shown that oocytes that have not yet achieved competence display RNA transcripts with shorter poly(A) tails than fully competent oocytes. These experiments point to a direct link between the extent of polyadenylation, mRNA stability, and the degree of competence during oocyte maturation and suggest the possibility to monitor polyadenylation as a marker of the female gamete competence (Fig. 3.2).

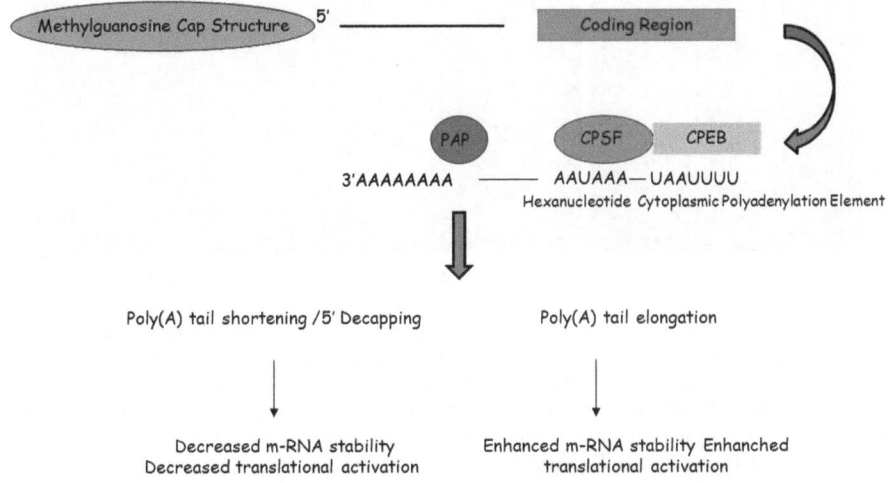

Fig. 3.2 Length of the poly(A) tail regulates maternal mRNA translation. Poly(A) tail is localized in the 3′ UTR region and controls the transcript stability based on its extension. In particular, oocyte mRNA with a short poly(A) tail is translationally inactive, while mRNAs displaying an extended poly(A) tail are stable and translationally active

A further regulatory system which is active in the oocyte and early embryo is the localization and sorting of organelles to distinct regions of the oocyte cytoplasm. This phenomenon, defined as cell streaming represents a powerful tool that allows a cell to distribute energy resources, to establish specific gradients, and to control gene expression, ensuring that the adequate concentrations of the protein are present where required.

A typical feature of the maturing oocyte is the relocation of mitochondria and this has emerged as a key event for the acquisition of competence. Changes in mitochondrial distribution are a faithful marker of cytoplasmic streaming and a subtle indicator of the oocyte capacity to support embryo development. In particular, it has been suggested that only oocytes possessing high developmental competence display mitochondria relocation ability. Three distinct distributions have been identified and correlated to different functional stages: peripheral (prevalent in the majority of the oocyte at the beginning of maturation), semiperipheral (distinctive of intermediate stages), and diffused (typical of fully mature gametes). Little or no relocation is detected in low-competence oocytes where mitochondria are left in a peripheral distribution. These observations suggest that a non-diffused distribution of mitochondria in the cytoplasm may be correlated to low developmental ability.

Similarly, the process of mRNA sorting and translation at specific areas within the cell has been described in several different experimental models and is accepted as an intriguing regulatory mechanism of meiotic maturation and early embryonic development (Fig. 3.4). mRNA sorting represents a key regulatory step of gene expression in non-transcribing cells and is one of the ways translation is locally regulated, ensuring appropriate concentrations of the encoded

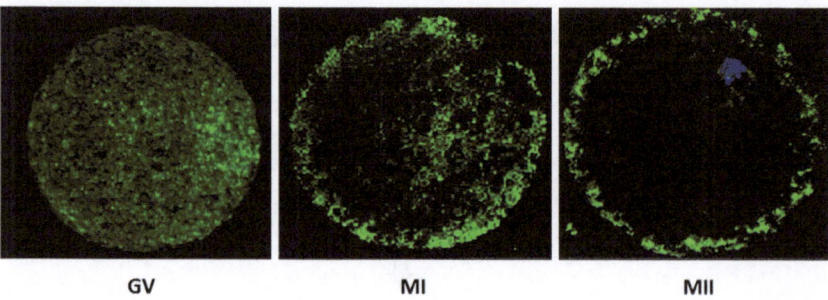

Fig. 3.3 Localization of Staufen in GV, MI, and MII oocytes. Localization is a fundamental regulatory system that ensures an adequate protein distribution to be active in the oocyte. Three distinct distributions have been identified for Staufen, an RNA-binding protein involved in RNA transport to the different areas of the cell. In particular, a diffuse distribution is visible in GV oocytes, while the signal becomes semiperipheral and granular in MI oocytes. Finally, a peripheral localization is present in fully mature oocytes (MII stage)

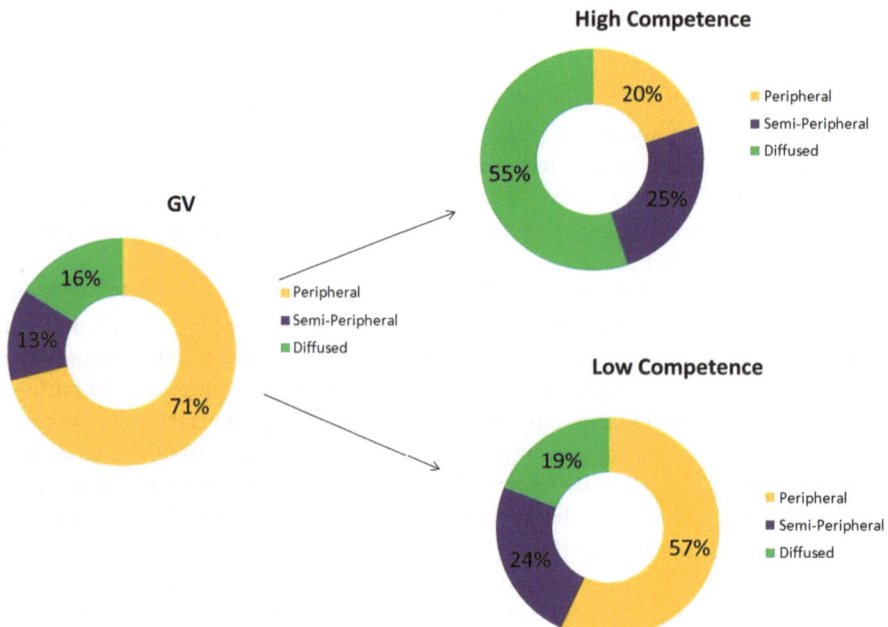

Fig. 3.4 mRNA localization in GV and in MII oocytes with high or low developmental competence. mRNA sorting and translation represent a key regulatory mechanism in the meiotic maturation of the female gamete and early embryonic development. GV oocytes present most of the mRNA in the peripheral area (71 %). During the maturation process, transcripts are relocated in the ooplasm. Indeed, high competence oocytes show a diffuse mRNA localization (55 %), whereas low competence ones display only the 19 % of mRNA diffused in the ooplasm and 57 % remaining confined to the peripheral area

protein where it is needed. RNA localization and translation at specific sites within the ooplasm is a property common to many species and is made possible through the interaction with specific RNA-binding proteins that are able to recognize mRNAs as a target for transport, to relocate them along the cytoskeletal tracks, and to anchor them at specific docking sites. Several molecules have been recently characterized as shuttling proteins, actively involved in RNA relocation to the translation spot. Among the many we will here mention Oskar, Staufen (see Fig. 3.3), Pumilio that are considered to be indispensable for the establishment of polarity in the oocytes of lower species, aimed at specifying a morphogenetic gradient for the determination of the embryonic axis (see morphogens and the French flag model). Their specific role in mammalian species is currently under investigation.

A successful transition from maternal to autonomous embryonic transcription is the most obvious step but it is at present fully accepted that oocyte influence extends well beyond this stage into embryogenesis. Embryonic development will not start in the absence of a fully competent oocyte. On the other hand, no options are available to restore gamete competence once it is affected.

Therefore, it comes with no surprise that development can progress to a great extent without any paternal contribution as demonstrated by parthenogenetic activation.

Further Reading

Brevini TA, Cillo F, Antonini S, Gandolfi F (2007 Mar) Cytoplasmic remodelling and the acquisition of developmental competence in pig oocytes. Anim Reprod Sci. 98(1–2):23–38

Kugler JM, Lasko P (2009) Localization, anchoring and translational control of oskar, gurken, bicoid and nanos mRNA during Drosophila oogenesis. Fly (Austin) 3(1):15–28

Lackner DH, Bahler J (2008) Translational control of gene expression: from transcripts to transcriptomes international review of cell and molecular biology, vol 271. Elsevier Inc

Latham KE, Schultz RM (2001) Embryonic genome activation. Front Biosci 1(6):D748–D759

Quenault T, Lithgow T, Traven A (2011 Feb) PUF proteins: repression, activation and mRNA localization. Trends Cell Biol 21(2):104–112

Santos F, Hendrich B, Reik W, Dean W (2002 Jan 1) Dynamic reprogramming of DNA methylation in the early mouse embryo. Dev Biol 241(1):172–182

Stutz A, Conne B, Huarte J, Gubler P, Völkel V, Flandin P, Vassalli JD (1998 Aug 15) Masking, unmasking, and regulated polyadenylation cooperate in the translational control of a dormant mRNA in mouse oocytes. Genes Dev 12(16):2535–2548

Van Blerkom J (2004 Sep) Mitochondria in human oogenesis and preimplantation embryogenesis: engines of metabolism, ionic regulation and developmental
competence. Reproduction 128(3):269–280

3.2 Oocyte as a Source of Uniparental Pluripotent Cells

Experiments carried out with parthenogenetic embryos have demonstrated that
mammalian oocytes are able to differentiate in the various tissues of the body
without the contribution of the male gamete. They are, however, limited by their
inability to generate a functional placenta and cannot develop to term. Because
of this, it is widely accepted that parthenotes arrest gestation around implantation
(see Table 1.4). Based on these observations, it is not unexpected that pluripotent
cell lines can be derived from embryos obtained after parthenogenetic activation
protocols. Much less understood is whether and how parthenogenetic pluripotent
cells can bypass the limits derived from imprinting and allow the development of a
full individual.

The first successful attempt to generate a cell line from an activated oocyte
dates back to over 20 years ago and was carried out in the mouse. However, it was
soon followed by the derivation of parthenogenetic cell lines in the monkey and
in rabbit, indicating the possibility to obtain pluripotent cell lines of uniparental
origin in different species (Fig. 3.5). The results were very encouraging because
it appeared that these cells displayed the main properties of normal biparental
embryonic stem cells.

They were stable in culture, they could be addressed in vitro to form embryoid bodies when cultured in suitable conditions and generated teratomas, when
injected in immunosuppressed mice. These results indicated the possibility to
derive parthenogenetic stem cells in most species, including also the human and
therefore suggested that supernumerary oocytes could represent a less contro-
versial source of embryonic stem cells to be used for therapeutic applications.
However, studies performed in mouse cell lines demonstrate that parthenogenetic
cells may be restricted in their differentiation ability. Injection of diploid mouse
parthenogenetic lines into normal blastocysts generated chimeras but the contribution to skeletal muscle and testis formation was considerably lower than for other
tissue tested. The restricted developmental potential of uniparental cells was further suggested by little skeletal muscle found in teratomas.

Another aspect that must be clarified is related to the fact that parthenotes
derive from single eggs. Although this is an advantage in order to generate his-
tocompatible stem cells, considerably reducing the complexity of tissue match-
ing for transplant purposes, it must be emphasized that the occurrence of a high
degree of homozygosity can represent a severe risk. Loss of heterozygosity may,
in fact, amplify negative genetic element possibly harboring in the genotype
and constitutes a crucial limit since it is regularly associated with chromosome

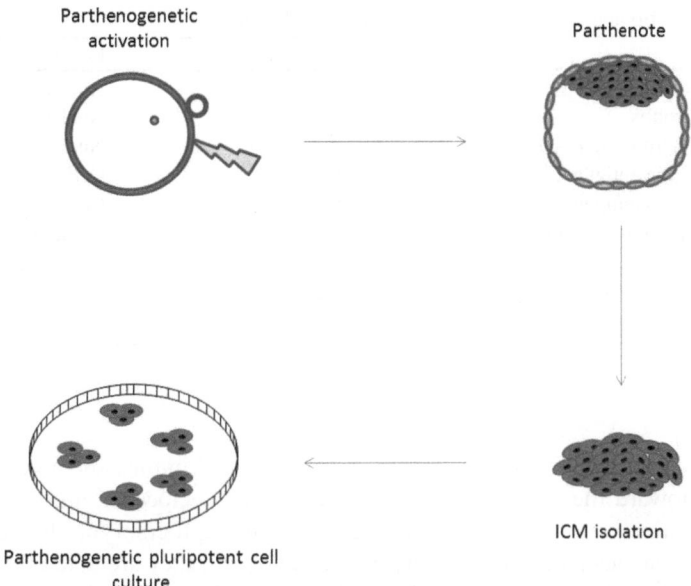

Fig. 3.5 Parthenogenetic embryonic stem cell isolation. A parthenogenetic embryo is cultured until the blastocyst stage, its inner cell mass (ICM) is mechanically/immunosurgically isolated and plated on mitotically inactivated feeder cells

instability and carcinogenesis in humans. It is important however to remember that, as more extensively discussed in the chapter "Oocyte activation", the choice of the activation protocol greatly affects the level of heterozygosity in the embryos obtained and, consequently in the lines generated from the latter. All these observations clearly point out that a crucial aspect in parthenogenesis is the choice of the activation procedure and further information are needed in order to clarify the gains and losses of the different technical approaches, estimating efficiency versus safety.

Human parthenogenetic cell lines display many of the features characteristic of biparental cells. They show self-renewal ability and can be easily propagated in vitro. They express cell markers that characterize pluripotency in human embryonic stem cells: *Oct-4, Nanog, Rex-1, Sox-2, alkaline phosphatase, SSEA-4, TRA 1-81* (see Table 3.1).

Similarly to embryonic stem cells, they show high telomerase activity when they are in an undifferentiated state, while they respond to induction toward the specification of a tissue specific lineage turning down this activity, indicating that a physiologically normal control of telomerase activity is present and active in cells of uniparental origin.

Parthenogenetic cells of different species are able to respond to special media formulations and, when grown in suspension, form embryoid bodies. These

Table 3.1 Pluripotency-related markers and their localization

Gene	Protein encoded	Localization
POU class 5 homeobox 1	Oct-4	Nucleus
Nanog homeobox	Nanog	Nucleus
SRY (sex determining region Y)-box 2	Sox-2	Nucleus
Tumor rejection antigen-1-60	Tra-1-60	Cell surface
Tumor rejection antigen 1-81	Tra-1-81	Cell surface
Zinc finger protein 42 homolog	Rex1	Nucleus
Alkaline phosphatase	Ap	Cytoplasm
Stage-specific embryonic Antigen-3	Ssea-3	Cell surface
Stage-specific embryonic Antigen-4	Ssea-4	Cell surface

three-dimensional clumps of cells rapidly progress and initiate spontaneous differentiation toward the three germlineages—endoderm, ectoderm, and mesoderm—that are distinctive of somatic cells (Fig. 3.6). Although embryoid bodies display non-homogeneous patterns of differentiated cell types, they are able of responding to developmental cues similar to those directing early embryo development. This property makes them highly suitable to investigate the mechanisms driving early differentiation in early embryos and, while obtained from parthenogenetic cells, they may be a valuable tool to elucidate plasticity of uniparental organisms and to study the specific contribution of the maternal genome.

Using both embryoid bodies and monolayer cultures, it has been shown that parthenogenetic cells respond to stimulation with Sonic Hedgehog and retinoic acid (molecules known to play a role in patterning the central nervous system) and have the ability to generate early and mature cell types of the neural lineage. Their asymmetric imprinting does not seem to affect cell ability to respond to developmental cues and appear to reproduce the induction process described for cells of biparental origin.

Likewise, exposure to specific cytokines and adequate culture conditions have demonstrated parthenogenetic cell ability of in vivo clonal colony formation. It has been further shown that this protocol address cells toward the hematopoietic lineage, with the formation of lymphoid, erythroid, and myeloid subpopulations.

These evidences were separately accumulated in different laboratories and indicate that the differentiation potential of uniparental cell lines may be less limited than that of parthenogenetic cells in chimeras (described above) and that these cells can represent a precious and informative tool for the investigation of the main and fundamental process driving cell lineage commitment and differentiation in vitro.

Injection of parthenogenetic cells in immunodeficient mice has been reported to generate contrasting results, conducting to either poor differentiation or to the formation of tumors. Interestingly however, the newly generated tumors are mainly related to abnormal muscular differentiation and suggest that a deregulated control of the mechanisms leading to proliferation/differentiation is possible in stem

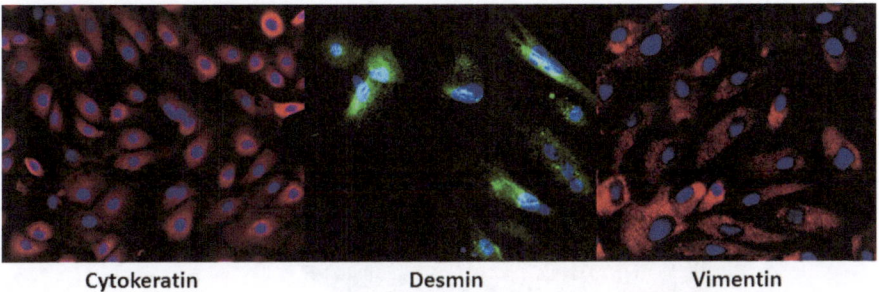

Cytokeratin   Desmin   Vimentin

Fig. 3.6 Parthenogenetic cells and differentiation. Parthenogenetic pluripotent cells are able to spontaneously differentiate and to respond to special media formulations, generating cells belonging to the three germ layers. Immunopositivity for cytokeratin, desmin, and vimentin demonstrates cell differentation ability toward ectoderm, mesoderm, and endoderm lineages, respectively

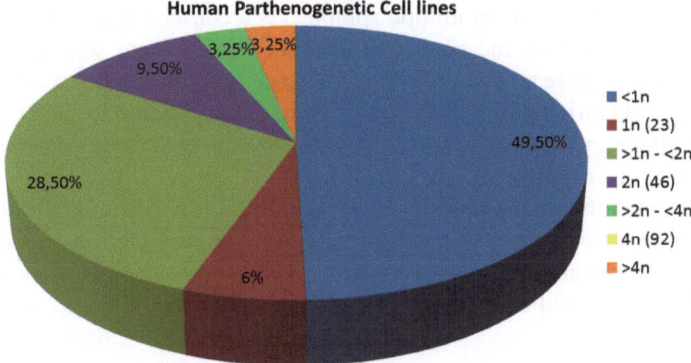

Fig. 3.7 Ploidy in human parthenogenetic cell lines. A high rate of aneuploidy is observed in parthenogenetic cells. In particular, hypohaploid (49, 50 %) and hypodiploid (28, 50 %) configurations are the most frequent setups detected in these cell lines

cells obtained through parthenogenesis. These tumors may find their explanation in the high incidence of aneuploidy reported in parthenogenetic embryos of different species. Karyotype analyses of sheep, pig, and bovine parthenotes have indeed demonstrated abnormal chromosomal complements. While the results available in the mouse are contradictory, data for human parthenotes are fairly limited but confirm a high incidence of aneuploidy in preimplantation embryos, derived from assisted reproduction procedures. Among the possible explanations, it has been suggested that the lack of the paternal centrosome, that is contributed by the sperm at fertilization, may represent a possible reason for the high rate of aneuploidy in parthenogenetic cells (Fig. 3.7).

It may be hypothesized that, in the absence of the sperm centriole, uniparental embryos are unable to correctly reaggregate the oocyte centrosomal material

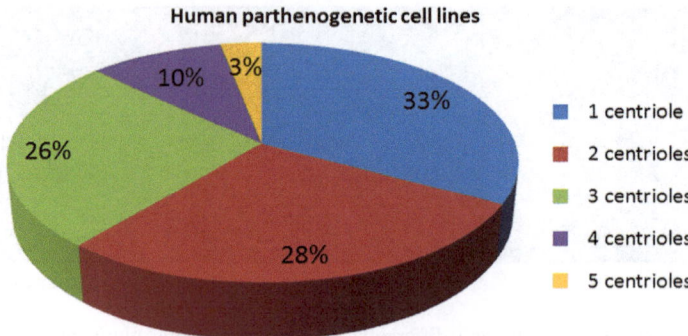

Fig. 3.8 Centriole pair number in human parthenogenetic cell lines. A high incidence of super-numerary pairs of centrioles was observed in human parthenogenetic cells. In particular, 39 % of cells display centriole number alterations

and form a properly arranged mitotic spindle. As a consequence, the formation of multipolar spindles with incorrect and unequal chromosome segregation during cell division. All these aspects are in agreement with the presence of an abnormal number of centrioles (Fig. 3.8) and with altered expression levels of molecules involved in the regulation of centriole formation spindle assembly detected in parthenogenetic cells of different species.

The information summarized in this chapter describe and stretch the concept that human parthenogenetic pluripotent cells can be generated, cultured and can differentiate in mature cell types in vitro, but also emphasize the risk of altered proliferation ability and chromosome instability in these cells.

Further Reading

Brevini TA, Gandolfi F (2008 Feb) Parthenotes as a source of embryonic stem cells. Cell Prolif 41(Suppl 1):20–30

Brevini TA, Pennarossa G, Antonini S, Paffoni A, Tettamanti G, Montemurro T, Radaelli E, Lazzari L, Rebulla P, Scanziani E, de Eguileor M, Benvenisty N, Ragni G, Gandolfi F (2009a) Cell lines derived from human parthenogenetic embryos can display aberrant centriole distribution and altered expression levels of mitotic spindle check-point transcripts. Stem Cell Rev 5(4):340–352

Brevini TA, Pennarossa G, Maffei S, Tettamanti G, Vanelli A, Isaac S, Eden A, Ledda S, de Eguileor M, Gandolfi F (2012) Centrosome amplification and chromosomal instability in human and animal parthenogenetic cell lines. Stem Cell Rev

Drukker M (2008 Apr) Recent advancements towards the derivation of immune-compatible patient-specific human embryonic stem cell lines. Semin Immunol 20(2):123–129

Müller R, Lengerke C (2009) Patient-specific pluripotent stem cells: promises and challenges. Nat Rev Endocrinol 5(4):195–203

3.3 Spindles of Uniparental Origin in Mammals

As described above, mammalian oocytes can be parthenogenetically activated using various physical or chemical stimuli and their maternal haploid genome can be diploidized. The relationship between parthenogenesis and centrosomes is complex and differences exist among the various species. Furthermore, it is important to consider that usually in non-rodent species the sperm contributes the proximal centriole that is introduced into the ooplasm together with the sperm head. Indeed, centrosomal material of MII oocyte does not organize into unified foci and is neither capable to generate astral microtubules nor a correctly oriented spindle in the absence of centrioles.

The majority of information present in the literature concerning the consequences of the absence of paternal centrioles on parthenote centrosome is derived from studies carried out in lower species.

At present, two basic mechanisms involved in the formation of the bipolar mitotic spindle are hypothesized:

1. A "centrosome-dominated" pathway, also known as early event, in which the assembly and organization of the astral spindle is directly driven by centrosomes;
2. A "chromosome-dominated" pathway, also known as late event, in which the assembly of anastral spindles is regulated by the chromosomes that organize microtubules.

In agreement with these observations, previous studies carried out in sea urchin eggs demonstrated that a treatment with ammonia induces parthenogenetic activation supporting the chromosome-dominated pathway, without the formation of de novo synthesized centrioles. In contrast, the use of standard activation protocols stimulates new centriole formation, indicating that eggs undergo both the early and late events of activation.

Furthermore, sea urchin together with insects and several other lower species are obligatory or facultative parthenotes. In these species, in fact, the oocyte generates a complete and functional centrosome and replenishes all components of the zygotic centrosome in the absence of the male gamete. Parthenogenetically activated eggs give origin to multiple cytoplasmic asters, containing centrosomal proteins and centriole, possibly due to the absence of a correct control along the process of spindle formation. Two of these multiple asters become stably associated with the female pronucleus and generate the mitotic spindle, whereas the others degenerate.

Furthermore, recent studies performed on sea urchin activated eggs, where paternal centrosomes were introduced, and on Drosophila activated and then fertilized oocytes, demonstrated the dominance of the "sperm-derived centriole-centrosome complex" over the ooplasmic centrosomal components. These events seem to play a fundamental role in embryonic development, stem cell division, and differentiation, and in the ontogenesis of several types of tumors.

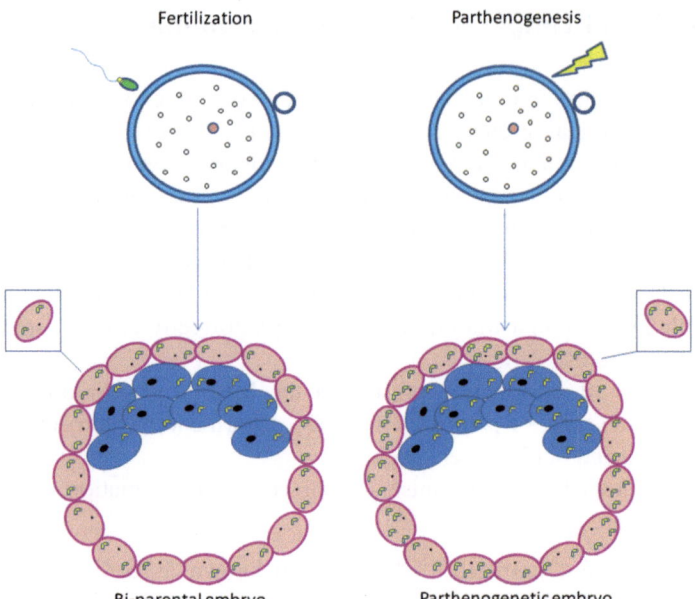

Fig. 3.9 Correlation between IVF/parthenogenesis and centriole number in non-rodent mammalian species. It is hypothesized that the absence of paternal centriole in parthenogenetic zygotes leads to a centrosome amplification process, causing the formation of supernumerary pairs of centrioles that are not able to clusterize. This is likely to be due to a deregulation of the centrosome-dominated pathway, involved in the formation of the mitotic spindle, or to the lack of appropriate centriole inhibition mechanisms

In mammalian species it is well known that male and female centrosomes undergo a reciprocal reduction during gametogenesis and only at the time of fertilization the centrosomal structure is restored and reorganized. By contrast, scattered and incomplete information about the link between parthenogenesis and mitotic spindle organization are available. Only recently has it been hypothesized that in different mammalian species, with the exception of the mouse oocyte that does not receive centrosomes from spermatozoa, standard activation protocols may stimulate both the early and late activation events, causing the formation of supernumerary pairs of centrioles that are not able to clusterize (Fig. 3.9).

At the same time, it is possible to assume that the lack of paternal contribution in parthenotes may produce an alteration in the negative regulatory mechanism, suggested to turn off centriole de novo assembly, and in the expression levels of several mitotic spindle-related molecules.

Indeed, several centrosomal proteins are closely related to centrosome organization and studies investigating possible alterations in their expression level are often used to understand or confirm cell centrosome defects (see Table 3.2).

Recent studies showed the presence of multipolar spindles together with altered expression levels for several mitotic spindle-related molecules in parthenogenetic embryos and derived cell lines of different species. These observations

Table 3.2 Centrosomal proteins, their localization, and effects of altered expression levels

Gene symbol	Gene name	Localization	Effect of expression level changes
AURKA	Aurora kinases A	Centrosome from the time of duplication until the end of mitosis	*Overexpression* block of spindle assembly checkpoint, arrest of mitosis, incomplete cytokinesis, centrosome amplification, multipolar spindles, multinucleation
AURKB	Aurora kinases B	Heterochromatin in early mitosis, in central spindle during anaphase, in cell cortex when the contractile ring forms and in midbody during cytokinesis	*Depletion* alterations of function and localization of the spindle checkpoint components, misaligned chromosomes, syntelic attachments of chromosomes to the spindle poles, cell division failure, endoreduplication
AURKC	Aurora kinases C	Centrosome from anaphase to telophase	Unknown
PLK1	Polo-like kinase 1	Centrosome	*Depletion* multipolar spindle formation, improper cytokinesis, apoptosis
PLK2	Polo-like kinase 2	Centrosome	*Overexpression* increase of centrosome numbers; *Depletion*: decrease of centrosome numbers
BUB1	Budding uninhibited by benzimidazoles 1	Kinetochores	*Depletion* reduction of tension across the centromere, spindle pole fragmentation, mixture or unaligned chromosomes, mitotic arrest
CENPE	Centromere protein E	Fibrous corona of the kinetochore	*Depletion* mono-oriented chromosomes, mitotic arrest, apoptosis
TTK	TTK protein kinase	Kinetochores	*Depletion* mitotic abnormalities, chromosome number alterations
MAD1	Mitotic arrest deficient-like 1	Kinetochores	*Depletion* unaligned chromosomes, mitotic arrest
MAD2	Mitotic arrest deficient-like 2	Kinetochores	*Depletion* monoastral mitoses, aberrant chromosome segregation, aneuploidy
CCNF	Cyclin F	Centrosomes	*Depletion* centrosomal and mitotic abnormalities, multipolar spindles and/or asymmetric bipolar spindles with lagging chromosomes

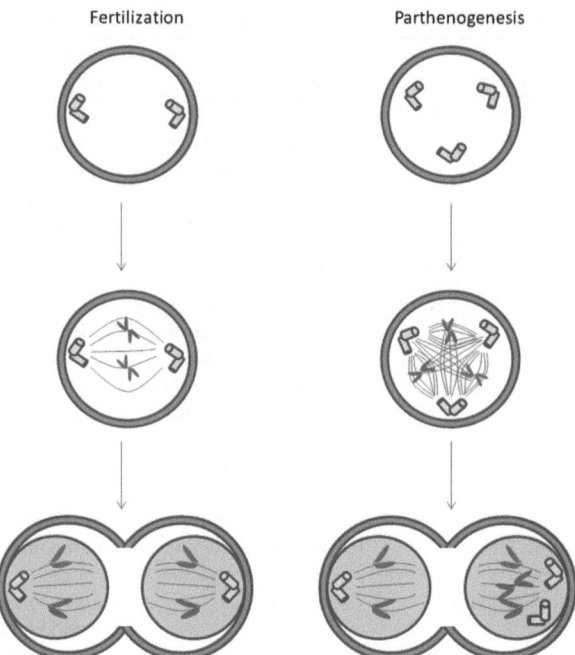

Fig. 3.10 Centriole number alteration in parthenogenetic embryos causes chromosome malseg-regation. In non-rodent mammalian species (*left panel*), the oocyte receives the proximal centriole from the sperm. This centriole replicates and, at syngamy, the duplicated set of organelles relo-cates to opposite poles to form the mitotic spindle poles. This arrangement drives the first embryo mitotic division with a correct chromosome segregation. Conversely, pathenogenetic embryos are characterized by the absence of the paternal centriole. This leads to the formation of supernumer-ary pairs of centrioles that are not able to clusterize, causing unequal chromosome separation

are compatible with the elevated incidence of aneuploidy detected in pre- and post-implantation parthenotes of several mammals. In all these species, in fact, the centriole, introduced by the sperm, contributes to reorganize oocyte centroso-mal material to form a fully functional zygote aster and mitotic apparatus. This suggests that centriole absence may cause an aberrant centrosome formation and an increased number of centrioles, leading to unequal chromosome segregation (Fig. 3.10).

Indeed, severe chromosomal alterations were identified when parthenogen-esis is induced in non-rodent mammalian oocytes. In particular, polyploidy, mix-oploidy, and aneuploidy were detected in bovine, porcine, ovine, and human embryos, irrespective of the activation protocol used (electric stimulation, etha-nol, ionomycin, puromycin, cytochalasin B, cytochalasin D, 6-DMAP). Moreover, chromosome malsegregations have also been described in spontaneously activated bovine and human oocytes, confirming that the lack of paternal contribution to the parthenote centrosome may be the possible cause of a high rate of aneuploidy.

Further Reading

Brevini TA, Pennarossa G, Antonini S, Paffoni A, Tettamanti G, Montemurro T, Radaelli E, Lazzari L, Rebulla P, Scanziani E, de Eguileor M, Benvenisty N, Ragni G, Gandolfi F (2009b) Cell lines derived from human parthenogenetic embryos can display aberrant centriole distribution and altered expression levels of mitotic spindle check-point transcripts. Stem Cell Rev 5(4):340–352

Brevini TA, Pennarossa G, Maffei S, Tettamanti G, Vanelli A, Isaac S, Eden A, Ledda S, de Eguileor M, Gandolfi F (2012) Centrosome amplification and chromosomal instability in human and animal parthenogenetic cell lines. Stem Cell Rev

Henson JH, Fried CA, McClellan MK, Ader J, Davis JE, Oldenbourg R, Simerly CR (2008 May) Bipolar, anastral spindle development in artificially activated sea urchin eggs. Dev Dyn 237(5):1348–1358

Schatten H, Schatten G, Mazia D, Balczon R, Simerly C (1986) Behavior of centrosomes during fertilization and cell division in mouse oocytes and in sea urchin eggs. Proc Natl Acad Sci U S A 83:105–109

Schatten H, Sun QY (2011 Dec) New insights into the role of centrosomes in mammalian fertilization and implications for ART. Reproduction 142(6):793–801

Schatten H, Walter M, Biessman H, Schatten G (1992) Activation of maternal centrosomes in unfertilized sea urchin eggs. Cell Motil Cytoskel 23:61–70

3.7 Stem Cells as a Source of Oocytes

A central dogma of mammalian reproductive biology is that females are born with a fixed, non-renewing pool of germ cells, arrested in the prophase of meiosis I (oocytes) and enclosed by somatic cells in follicles. The number of oocytes decline throughout postnatal life through mechanisms that involves apoptotic process, leaving the ovaries devoid of germ cells in due time. This is one of the recognized causes that limit female fertility with age through the exhaustion of the oocyte reserve during the years. The limited and finite number of fully grown oocytes is also the main limit to the application of somatic cell nuclear transfer for the generation of immunologically matched stem cells or to the implementation of assisted reproduction programs for the conservation of endangered species.

However, in recent years, different experimental approaches as well as unexpected findings are providing a completely new perspective and the promise of an infinite source of oocytes.

Embryonic stem cells (ESCs) injected into a blastocyst have the ability to contribute not only to any somatic tissue but also to the germline.

When pluripotency-maintaining factors are withdrawn from the culture media and/or the feeder layer is removed, ESCs quickly differentiate into a heterogeneous mixture of cell types representing the three somatic germ layers. It is reasonable to assume that ESCs may be induced to differentiate into germ cells (Fig. 3.11).

Reports using mouse ESCs have described oocyte-like and sperm-like differentiation, or both, as well as successful experiments that allow for primordial germ cell-like development. In contrast, studies carried out in the human ESCs reported both the spontaneous and induced differentiation of germ cells but at present, it has not been possible to obtain more mature gametes.

Hubner et al. were the first to develop a strategy for mouse ESCs differentiation into oocytes using an ESC line genetically modified with a germ cell-specific Oct4 promoter driving a GFR reporter construct. In appropriate culture conditions, ESCs differentiated into c-KIT/GFP positive cells that expressed low levels of VASA and that are likely to represent germ cells through their migratory stage. Further, in vitro culture induced down-regulation of c-KIT together with an increase of VASA expression (as expected in post-migratory germ cells).

After 16 days of culture the meiotic marker SCP3 was found in the nucleus of oocyte-like cells but no distinct chromosomal alignment was detectable. Irrespective of this abnormality, the culture for further 10 days has been described to lead to the formation of follicle-like structures that released oocytes of 50–70 μm and express some specific markers like Zp2/3 and Figla. Furthermore, and even more surprisingly, a few days later preimplantation embryos were observed, which were likely to be the result of a spontaneous parthenogenetic activation. Further studies carried out using the same protocol conducted to the formation of follicle-like aggregates capable to secrete high levels of estrogens into the medium. These structures, however, were unable to progress beyond an abnormal meiotic prophase.

Recent attempts were based on a two-step strategy which included an initial phase where ESCs were induced to differentiate into germ cells with retinoic acid, followed by the coculture of the resulting EBs with granulosa cells. This allowed the formation of oocyte resembling cells with a diameter of up to 25 μm expressing specific genes like *Vasa*, *Scp3*, and *Gdf9* but that showed no sign of follicular organization.

Possibly, the most striking results were generated in experiments that obtained both types of gametes from male mouse ES cells without any genetic manipulation or preselection. The differentiation of both gametes was induced by retinoic acid within non-adherent EBs and was demonstrated by the expression of early and late germ cell-specific genes in the correct order. At the end of the culture period, even a putative blastocyst-like structure was observed but it was not possible to determine whether it was originated by fertilization or parthenogenetic activation.

Even if these results are very encouraging it must be noted that oocyte-like cell maturation, oocyte functionality or their ability to be fertilized and produce offspring was not demonstrated in any of the above reports. A normal progression through meiotic prophase I and/or meiotic arrest seems to be beyond the reach of the current methods.

The possibility to obtain gametes from human ESCs was suggested by the observation that a subpopulation of undifferentiated human ESCs express markers

Fig. 3.11 Events and molecular keys involved in gamete differentiation from stem cells

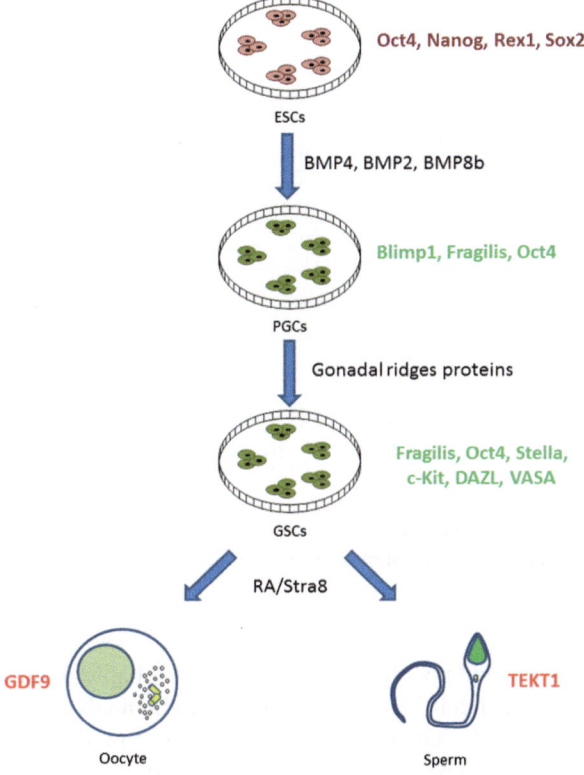

common to both inner cell mass cells and germ cells like *OCT4*, *STELLAR*, and *NANOS1* as well as the gonocyte-specific *DAZL* (Fig. 3.12). Further differentiation into EB determined a down-regulation of these genes accompanied by an increased expression of *VASA*, a later stage marker of germ cell differentiation. Several studies have described that, upon various intervals of EB culture, with or without specific inducing factors like BMP, it was possible to observe the expression of the oocyte specific marker *GDF9* and the meiotic markers *SCP1* and *SCP3* but, similarly to the mouse, chromosomal alignment indicative of meiotic prophase I progression was not observed.

It is worth to notice that only a limited number of ES cells become PGCs in culture. This is consistent with the situation present in normal development, where a limited number of the cells in the proximal epiblast are allowed to differentiate into germ cells. It is believed that the "choice" of germ cell fate may depend on differences among the ES cells, and the choice is then reinforced by interactions between the cells becoming germ cells and those retaining the somatic cell fate.

A promising new protocol based on cell selection by FACS and that allows to differentiate human embryonic stem cell lines, maximizing the numbers of primordial germ cells has been recently described. The cells obtained in this way have high-level expression of germ cell-specific *VASA*, *SCP1*, and *SCP3* genes while

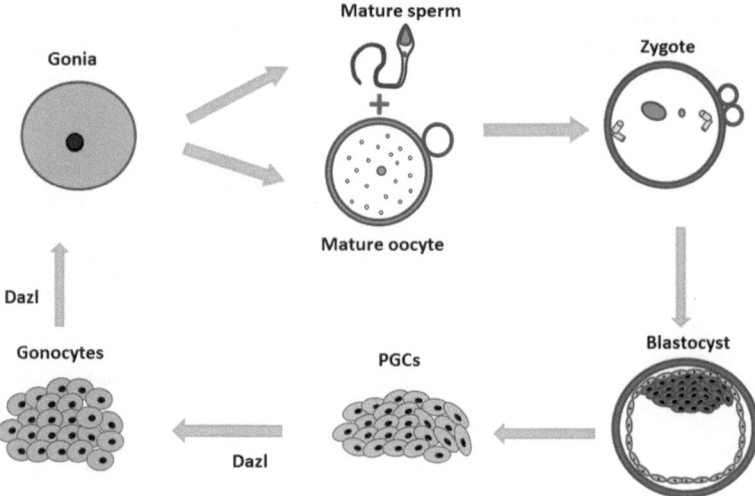

Fig. 3.12 Differentiation of gametes from stem cells and the role of DAZL in both imprint erasure and establishment of sex specific imprints

a small subpopulation appeared to be haploid further supporting their primordial germ cell identity. Furthermore, the analysis of methylation patterns showed a series of changes suggesting that these putative hESC-derived primordial germ cells may have begun the epigenetic reprogramming process typical of these cells in vivo. The lower efficiency of human ESCs to differentiate into putative germ cells and their current inability to progress toward more mature cell types is likely to reflect the longer physiologic interval of human gamete differentiation compared to the mouse. Furthermore, it must also be reminded that human ESCs are not the equivalent of mouse ESCs but rather the equivalent of mouse epiblast stem cells (EpiSCs). The difference is likely to be very relevant since, in contrast to the mouse ESCs, mouse EpiSCs do not express germ cell markers, suggesting a non-germ cell origin.

Even if ESCs are the most likely candidates as a source of oocytes, data are available indicating that tissue-specific stem cells may be a possible source as well. In particular, fetal porcine skin stem cells and adult rat pancreatic stem cells were reported to have the ability to differentiate into oocyte-like cells in follicle-like aggregates that expressed germ cell markers. The logical and desirable evolution of these studies would be the differentiation of oocytes from induced of pluripotent stem cells (iPSCs) derived from any adult tissue of any individual.

The first results toward this ambitious goal have recently been published in a study which described the differentiation of human iPSCs into primordial germ cells (PGCs), obtained by coculture with human fetal gonadal stromal cells. Gene expression analysis and bisulfite sequencing determined that these cells correspond to committed first trimester germ cells, although the differentiation

Fig. 3.13 Schematic represen-
tation of the steps involved in
the derivation of iPSCs from
fibroblasts and their in vitro differentiation to oocytes or sperm
cells

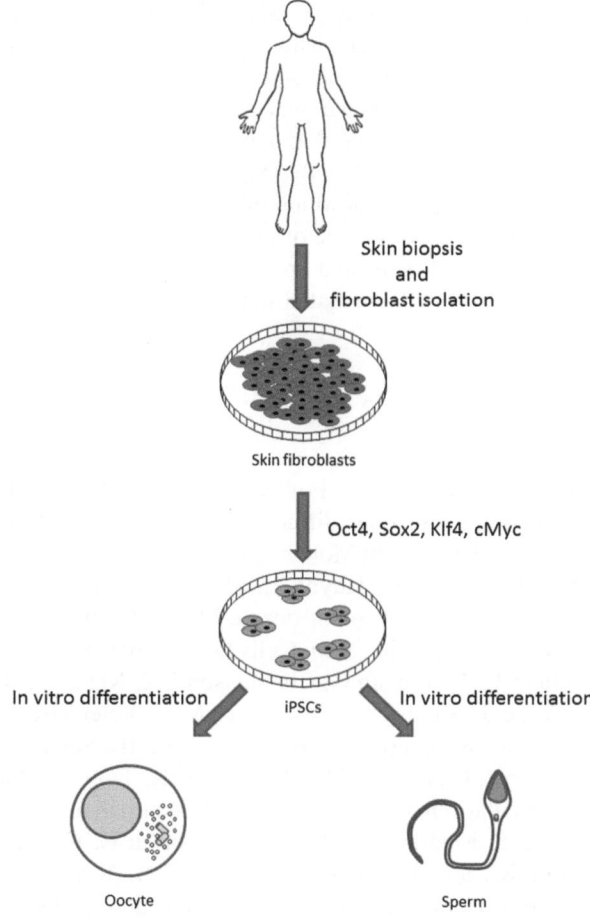

efficiency of iPSC was lower and less complete than that of normal hESC
(Fig. 3.13).

All the studies discussed above are based on the differentiation of stem cells
of different origin with no suggestion that such phenomena can happen physiologically. On the contrary, a series of recent studies revived the idea of continuous
oogenesis in postnatal ovaries by showing regeneration of oocytes from putative
germ cells in bone marrow and peripheral blood. Following the initial observation
that the quantitative analysis of healthy and atretic follicles in ovaries sterilized
with chemotherapy showed almost no difference in follicle count between control
and treated ovaries in 2 months, it has been shown that, bone marrow transplan-
tation in mice chemo-ablated with combined chemotherapy that kills all existing
germ cells restored oocyte production. However, since all offsprings were derived
from the donor germline mice the Authors suggested that the most likely mechanism of action of BM-derived cells is to reinstate recipient oogenesis.

These results as well as the idea itself that adult oogenesis actually exists has stirred a heated debate which is currently still ongoing. A further question we need to answer should postnatal oogenesis be confirmed, is related to the possible sources of oocytes. Is the bone marrow the only one in adult life? A strong alternative candidate as a possible source of oocytes has been described in the form of ovarian stem cells (OSCs) isolated from the ovarian surface. These cells appear to have the capacity of totipotent germline-competent embryonic stem cells. They were shown to be capable of differentiating into oocytes, fibroblasts, and epithelial and neural cell types and are known to have the capability of self-renewal, maintaining their undifferentiated state.

Further Reading

Clark AT, Bodnar MS, Fox M, Rodriquez RT, Abeyta MJ, Firpo MT, Pera RA (2004) Spontaneous differentiation of germ cells from human embryonic stem cells in vitro. Hum Mol Genet 13:727–739

Dyce PW, Shen W, Huynh E, Shao H, Villagómez DA, Kidder GM, King WA, Li J (2011) Analysis of oocyte-like cells differentiated from porcine fetal skin-derived stem cells. Stem Cells Dev 20:809–819

Hübner K, Fuhrmann G, Christenson LK, Kehler J, Reinbold R, De La Fuente R, Wood J, Strauss JF 3rd, Boiani M, Schöler HR (2003 May 23) Derivation of oocytes from mouse embryonic stem cells. Science 300(5623):1251–1256

Johnson J, Canning J, Kaneko T, Pru JK, Tilly JL (2004) Germline stem cells and follicular renewal in the postnatal mammalian ovary. Nature 428:145–150

Panula S, Medrano JV, Kee K, Bergström R, Nguyen HN, Byers B, Wilson KD, Wu JC, Simon C, Hovatta O, Reijo Pera RA (2011) Human germ cell differentiation from fetal- and adult-derived induced pluripotent stem cells. Hum Mol Genet 20:752–762

Song SH, Kumar BM, Kang EJ, Lee YM, Kim TH, Ock SA, Lee SL, Jeon BG, Rho GJ (2011) Characterization of porcine multipotent stem/stromal cells derived from skin, adipose, and ovarian tissues and their differentiation in vitro into putative oocyte-like cells. Stem Cells Dev 20:1359–1370

About the Authors

Tiziana A. L. Brevini is an Associate Professor of Anatomy and Embryology at the University of Milan, Italy. She graduated in 1989 and spent 3 years at the Department of Molecular Embryology, Cambridge (UK). She obtained a PhD in 1994 and then carried out research programs at Monash University, Melbourne and at the University of Adelaide, Australia. Her main area of research is addressed to the understanding of cell differentiation process and pluripotency related networks in mammalian cells and embryos.

Georgia Pennarossa is lecturer at the University of Milan, Italy. Graduated in 2007, she obtained a PhD in Biotechnology in 2012. She has been actively involved in research for the last 5 years at the Unit of Biomedical Embryology, University of Milan, Italy. Her scientific interest is addressed to the understanding of the main aspects related to gametogenesis, embryo development, and cell differentiation.

T. A. L. Brevini and G. Pennarossa, *Gametogenesis, Early Embryo Development, and Stem Cell Derivation*, SpringerBriefs in Stem Cells, DOI: 10.1007/978-1-4614-5532-5, © The Author(s) 2013

About the Authors

Index

T. A. L. Brevini and G. Pennarossa, *Gametogenesis, Early Embryo Development, and Stem Cell Derivation*, SpringerBriefs in Stem Cells, DOI: 10.1007/978-1-4614-5532-5, © The Author(s) 2013